AF131132

Radio's
LISTENING GROUPS

Radio's
LISTENING GROUPS

The United States and Great Britain

FRANK ERNEST HILL

&

W. E. WILLIAMS

NEW YORK: MORNINGSIDE HEIGHTS

COLUMBIA UNIVERSITY PRESS

1941

INTRODUCTION

A S ONE who is familiar with the work of both of the men who
have prepared the material presented in this book I consider
it an unusual privilege to stand off at a short distance and view
their separate analyses of an adult educational mechanism still in
the embryonic stage. Indeed, as the reader will discover soon after
he breaks into the subject matter, there are many reasons to suspect that the butterfly may never emerge from the cocoon, or if it
does, that it will not fly very far.

Nevertheless, in Europe, both in Britain and on the continent,
and in America as well, there has been enough perspiration over
listening groups to warrant diagnosis of the particular form of
educational pathology they might disclose. It is obvious that the
patient was still only under observation when Mr. Williams and
Mr. Hill used their scalpels. Now, with the onrush of the pan-
demic of war, it is unlikely that the symptoms they discovered and
analyzed will ever recur again in the particular form they found.

In spite of this set of conditions, I believe no one will aver that
what is written herein will ever cease to be of value. On the con-
trary, even if for mere historical record, when general peace again
comes to the world, and democratic processes of discussion and
group argument regain their rightful place, from the circumstances set down in these pages it will be possible to proceed in
ordered fashion to more advanced positions and with more logical
progress than would be possible had this inquiry not been undertaken in the period immediately preceding the outbreak of World
War II.

Because of the unique factors surrounding the growth of
American radio broadcasting, experimentation has been the order
of the day. Perhaps there has been too much of it—too much

broadcasting and too much experimentation. The mere freedom of the air in this country has led to a variety of outpourings from our millions of loudspeakers that sometimes nauseate the listener. However, he is forced to remember that just across the street, or around the corner, is another listener who not only does not share his judgment about a particular program but actually enjoys it and wants more of it. So that if in disgust listener No. 1 turns his dial, he ought to remember that listener No. 2 is having a grand time with himself and his family and his friends, drinking in and sopping up the "swing" or "education" or "horse opera" or "classical stuff" that raises the first listener's ire. In a country where variety of taste is so apparent, and where territorial reaches extend at such great lengths over such wide terrain, it may be that for experimental purposes alone, for a while at least, perhaps we cannot have too much broadcasting. And if that is so, by all means, let's have more experimentation, until we develop a mature sense of just what is a national norm and a reasonably sure standard of national taste at various levels of program selection.

Therefore, in respect to one mechanism which can be useful in the so-called "educational" variety of broadcasting, there should be every encouragement to test the efficacy of group listening. From Mr. Hill's findings it is obvious that there is little necessity for much stimulation in this country, granted normal conditions. On the other side of the water what will grow out of the post-war period no one can say. From both reports however it is clear that it is futile to expect that listening in groups will ever be susceptible of large independent development in spite of Mr. Williams' statement that it offers "a far higher standard of authoritative exposition than by any other mode of [adult] education." At best its value will be a supplemental one—a "secondary method," to use the British terminology.

It is rather surprising to find that in Britain there was little development of group listening in rural areas, and this discovery can well give American broadcasters pause. For we, in our characteristic enthusiasm have imagined that one of the greatest advantages our broadcasting can possibly achieve is to help the districts outside our industrialized areas. Britain's experience in this direction has been an eye-opener, and it may save us an unconscionable amount of hard labor and headache. Of course we must remember that we have enormously larger territory to cover than our cousins across the ocean and in a tight little island centralization of authority to some extent is forced, whether the victims or the beneficiaries of that condition recognize or admit the fact. Similar conditions would be impossible here.

To my mind the most striking statement by either author is the one Mr. Williams makes when he notes, as something perfectly obvious already, the "reaction against that glorification of discussion which has become a fetish of adult education." How deadly right he is! For in no country in the world could discussion for discussion's sake be more futile than we have witnessed it in America for the last decade. All the two- or three-way talk in the world, on the many issues that have been thrown in our faces apparently has not for an instant increased the capacity of this nation to think clearly or to act dispassionately. We still vote so that grass will not grow in the streets and almost before Norway was invaded we organized corps of women to practice shooting at descending parachutists. We condone graft and dishonesty in high places when the record thereof is clear so that anyone who runs may read. Merely providing opportunity for discussion, whether by groups listening to the radio, or by any other means, will never cure this inherent underdeveloped characteristic in our body politic. Something more fundamental is required, here and abroad. And that, perhaps, is the chief value to be gleaned from the wealth of informative material these two capable gentlemen have provided in the pages that follow.

Read Mr. Williams' Epilogue on page 256, and THINK.

LEVERING TYSON

Muhlenberg College,
Allentown, Pennsylvania
April 1, 1941

CONTENTS

PART I

THE UNITED STATES

BY FRANK ERNEST HILL

FOREWORD

Problems and Methods of an Educational Detective

THE PAGES which follow describe the first attempt to gather extensive information about radio listening groups in this country. The task proved to be in its quiet way a piece of pioneering, both in radio broadcasting and in adult education. Some readers will have a special interest in the methods used and the difficulties encountered; this foreword is written chiefly for them. Those who merely want to know what was discovered can turn to Chapter I and begin there, although these preliminary remarks will give them a framework for reading the rest of the book, and at times may even amuse them.

When the Study was begun in the spring of 1939, little dependable information existed as to groups of persons in America who gathered together to listen to radio programs. With a few exceptions, such bodies were lost in the invisible mass of our one hundred million radio listeners. The chances for locating them seemed little better than those of an ornithologist trying to determine the number of banded birds among the millions of wild ducks that visit American lakes, bays, and rivers.

In contrast, such groups in Great Britain were easy to study. With apologies for using so humble a figure, the British groups were as accessible as registered domestic animals in their stables or kennels. They were regularly counted and their sets were serviced by the British Broadcasting Corporation. Once authorized to study them, an investigator would know exactly where to find them, and would have difficulty only as he encountered suspicion, inertia, or that tendency to disappear under the examiner's nose which many listening groups suddenly develop. American groups, if they are to be included in the zoological simile used above, much more resembled the native fauna of a land—to be tracked down in a variety of regions by the persistent and sometimes baffled explorer, to be observed often only after a number of forays, and described and tabulated with a sigh of exhausted relief.

The complexity of the American adventure was recognized when it was planned. From the first there was no intention of making a comprehensive study. The numbers and "social significance" of such groups as existed were unknown. It was decided to make only a preliminary survey which would have an immediate practical value for persons working with, or wanting to work with, groups, and which would clear the ground for a more extensive examination if one seemed to be needed.

The objectives of the Study were: (1) to get some reasonable knowledge of the number and kinds of listening groups now active in the United States; (2) to visit some representative groups, observing them in action and talking with their leaders and those who might assist in promoting them; and (3) to get direct written testimony from a larger number of group leaders and members than those actually visited. Such testimony would deal with the respective organizations and procedures of groups; with their use of study aids; and with the defects and benefits found both in the radio programs which they heard and in their own activity as listeners.

A Crude Census of Listening Groups

To realize the first of these objectives—that is, to determine the prevalence and varieties of groups—it seemed desirable to make a systematic effort to locate them. Such an effort was also essential to the second and third objectives—that is, visiting and the gathering of direct testimony.

This effort to get reports on the numbers and locations of groups took several forms. One was the sending out of the inevitable questionnaire. Three questionnaires, prepared by the director, were reproduced and mailed by the U. S. Office of Education at Washington, D. C. One of these went to all of the 772 American radio stations then authorized to operate, another to all the colleges and junior colleges in the United States, and a third to a selected list of 200 American libraries.

In addition, the educational director of the Civilian Conservation Corps obligingly agreed, for his own information as well as for the benefit of the Study, to send out a questionnaire to educational advisers in all the CCC camps in the country; while, with the approval of the National Youth Administration's Washington Office, all state directors were asked for data on listening groups at NYA centers.

The Workers Education Bureau also sent out a questionnaire to American labor councils with the idea of discovering possible similar group activity in organized labor.

As the Federal Radio Education Committee's project on the Evaluation of School Broadcasts at Ohio State University was already dealing with classroom broadcasting, and as a time limit was necessarily fixed for the Study, no effort was made to locate listening groups in elementary schools or high schools.

All the questionnaires were simple: that sent to radio stations consisted of but four questions; the longest, sent to colleges and universities, contained but fifteen. The chief purpose of all the questionnaires was to discover what groups were definitely known to be functioning, and to get names and addresses of group leaders. The results were considered to be satisfactory. Altogether, 368 of 772 radio stations replied, 672 of 1,097 colleges, and 125 of 200 libraries. The answers seemed to give a general idea as to the prevalence of listening groups, and they furnished an ample number of "leads" through which the Study could get in touch with those representing group activity.

Meanwhile a second type of inquiry had gone forward. This consisted in a canvass, either by personal visit or by letter, of individuals familiar with broadcasting conditions, of a limited canvass of national organizations engaged in broadcasting, directors of various educational or semieducational radio programs, and heads of various governmental units with radio responsibilities. Such informal efforts proved to be fruitful, particularly with such organizations as Town Hall, Inc. (in charge of *America's Town Meeting of the Air*); the Family Life Institute under Dr. Alice Sowers at the University of Oklahoma (*Family Life Radio*

Forum); the State College at Ames, Iowa, and the University of Iowa at Iowa City, coöperating in the production of the *Radio Child Study Club;* the New Jersey State College of Agriculture at Rutgers University, New Brunswick, N. J., responsible for *Homemakers' Forum;* and the California Congress of Parents and Teachers, which produces the *Family Life Series.* However, reference to the activities of such bodies was usually made by radio stations in their replies to the questionnaires sent to them. Thus the second type of canvass served chiefly as a check upon the first.

Testimony from the Groups

To gather direct testimony from leaders and members of groups, three questionnaires were prepared: one for leaders, one for members, and one for registered listeners.[1]

All three questionnaires were formidable. That for group leaders contained 45 questions, that for registered listeners the same number, and that for group members 59. Shorter questionnaires obviously would have been easier to fill out, and hence the proportion of returns might have been larger. However, a questionnaire of fewer than twenty questions seemed to be superficial for our purposes, and if more questions were to be included, a fairly thorough coverage seemed worth risking. I was encouraged by Frank M. Stanton of the Columbia Broadcasting System to make the questionnaires full, his conviction being that if the questions were clear and simple, those answering them would reply to a large number as readily as to a small one. In practice this conviction seemed to be clearly supported in the case of two of the three questionnaires. Were I to reframe the questionnaire for members of groups, I should keep it within forty-five questions, like the two others.

The distribution and return of these three questionnaires was felt to be a highly important task—indeed, to be the heart of the study. It proved to be the most difficult work that was undertaken.

Originally the intention was to assemble completed question-

[1] The case of the registered listener and groups of registered listeners is discussed in Chapter I. Registered listeners are persons who write to a station or to the headquarters of those who produce a radio program, in order to get study aids or other materials which will help them listen more profitably.

naires from at least one hundred leaders; from one thousand members of listening groups; and from five hundred registered listeners. These questionnaires would represent a body of testimony from group listeners in America such as had never before been assembled. This testimony was expected to provide a number of interesting facts about how groups operated, and what they got or hoped to get from education by radio.

In the case of leaders and members, it was clearly necessary to find specific groups and to enlist their coöperation. As to the registered listeners it was necessary to get lists of such persons and in some fashion to get a questionnaire into the hands of each. The actual method used in the latter case was to send out the questionnaire with an explanatory letter and a stamped envelope for its return when completed.

At first it seemed as if the registered listeners might present a greater problem in securing information than would the groups. Even when lists were obtained from radio stations or program headquarters, how many individuals would fill out a forty-five-question document descending upon them from a strange if apparently responsible agency? If only ten percent, then five thousand persons would need to be circularized. In contrast, the groups seemed easier to handle. Once located, the group leader, it was hoped, would act as the representative for from ten to one hundred persons. The group would commit itself to coöperation by letter in advance of the receipt of the questionnaires. A hundred obliging groups, or perhaps even fifty (with an average of twenty members replying) would complete the quota.

Matters did not work out as anticipated. In practice, the registered listener proved to be the questionnaire-distributor's dream of a perfect collaborator, while the groups were often as difficult as reluctant witnesses at a racketeer's trial!

Adventures with Questionnaires

Early in the course of the Study a list of 376 persons was supplied to the director by the officials of Station WOSU at Ohio State University, all these individuals having been registered listeners to a French course broadcast from the station. It was thought that from 50 to 70 of these might reply, so the entire list

was used. The return was 247, or almost 66 percent! Eventually
10 lists of groups of registered listeners were obtained; more
could easily have been procured. The WOSU list proved to be
fairly typical. Of 830 persons addressed, 446 returned completed
questionnaires, or about 54 percent. One group of 27 persons
hearing the *Camera Club* program of Station KNX in Los
Angeles returned 21 questionnaires! The total response was impressive evidence of the interest and energy of individuals listening alone, and trying to study or get information by radio with
the assistance of printed or mimeographed aids.

In contrast, the groups of persons meeting together to hear
radio programs were pretty difficult to deal with.

In the first place, to locate a satisfactory variety of groups
was not easy. A thousand listeners from the many hundreds of
groups known to the officials of Town Hall, Inc., could perhaps
have been persuaded to fill out questionnaires. However, all
would have been group listeners to one program, and so large
a number from *America's Town Meeting of the Air* was not
desired. Town Meeting groups were kept to less than 29 percent
of all the groups from which testimony was gathered. Similarly,
an effort was made not to coöperate with too many groups listening to any one of the various programs dealing with family life
problems, although in the end about 47 percent of the groups
heard from, representing almost as many communities and seven
different series, were of this type.

Working with this attempt to get answers from group members of various types and in different parts of the country, those
making the study often found that a "lead" given to a group
proved to lead nowhere. A total of 219 letters written to possible
group leaders, brought no reply whatever. Other letters brought
answers such as, "No, I am not the leader of a radio listening
group," or "Yes, we had a group but it is now inactive," or "We
have talked of starting a group, but as yet have not organized."
Often a letter of inquiry revealed that the person addressed did
not have authority or information, but could refer the Study to
other persons who might give definite assistance. In such cases
two, three, or more letters were sometimes needed to bring actual
contact with a group or groups. Thus in the case of the California

Congress of Parents and Teachers, the president referred the director of the Study to the radio chairman. She in turn referred him to two district chairmen. One of these officials furnished the names and addresses of council chairmen, who finally sent in the names and addresses of groups! The other district chairman took charge—after a delay due to hospitalization—and personally distributed questionnaires. This particular adventure took from October to February, and threatened to leave the Study without returns from California P.T.A.'s. However, appeals for prompt action—tactfully supported by others to state pride—at length produced excellent results. In the final representation California led the 22 states which furnished groups, with a total of 25.

Once located, a group usually promised coöperation. A few leaders were prompt and thorough in carrying out their promises. Unfortunately they were exceptions. Enthusiastic requests for 10, 15, 25, and even 100 questionnaires were received relatively early in the course of the Study. An impressive number of pledges thus quickly piled up, but often these proved as worthless as checks written by a cheerful megalomaniac. Altogether but 24 percent of the questionnaires requested by groups and mailed to them were returned filled out.

Why Didn't They Answer?

The trouble may have lain partly in the length of the questionnaire for members. Some leaders were perhaps daunted by the idea of asking their group members to answer a questionnaire of fifty-nine questions. Some members may have noted the formidable character of the questionnaire, and put off dealing with it; others may have started filling it out and never finished. The idea of fifty-nine questions was I think a deterring element with some of them.

Yet the high percentage of responses to the registered listener questionnaire, which appeared to be even bulkier than the member questionnaire, being mimeographed while the latter was printed, made it doubtful if length were the chief cause for the relatively poor return. Except for a few additional questions about reading, discussion, and the social aspect of group listening, the two questionnaires were practically identical. Member questionnaires were filled out with no apparent difficulty by many who had no more than an elementary school education; by NYA students, both Negro and white; by one group of high school students; and by numerous other types of group members. A more probable cause of difficulty than length seemed to be the character of the groups and the role played by the leader in handling the questionnaires.

With rare exceptions, groups meet weekly or less often, and only at group meetings would the members in most cases have an opportunity to be told about the questionnaire. The leader would have to explain its character and use, and perhaps did not always take sufficient trouble to understand these before trying to make his presentation. Some leaders reported frankly that they had neglected to take the questionnaires to the first meeting after they had received them, and had then forgotten about their promised task of explanation and distribution until reminded by a follow-up letter. Others stated that they had been waiting for a well-attended meeting, and had not yet had one! Still others wrote that their members were suspicious of the purpose of the questionnaire. Some members, for example, objected to questions about age, religion, and politics; and in one case it was even suggested that the question about their automobiles—designed to reveal economic status—indicated that some manufacturer of motor cars had promoted the entire inquiry!

In addition, it is probable that some group members are attracted to groups on account of the novelty of the meetings and the social pleasure which these give, and are not greatly interested in any educational effort involving outside work. At least, group members seemed in general to be less serious in their interest than registered listeners, a number of whom wrote on their questionnaires, "I have enjoyed filling this out," or "You are doing a wonderful work!" or even, "Have you a questionnaire on musical (or some other) programs? I should like to answer that!"

Finally, the group member did not receive his questionnaire direct from our office in most cases, but indirectly. When the director of the Study attended group meetings, told about the purpose of the questionnaire, and distributed copies himself, as he took pains to do in the later stages of the work, the returns were about 70 percent. They also increased when, instead of having the leader collect as well as distribute the questionnaires, the Study merely asked him to distribute them, giving out a stamped envelope with each for its return.

Indeed, an increasing knowledge of how to deal with the groups brought increasing returns, and had there been more time, the full quota of 1,000 could have been obtained. But the necessity for tabulating and for bringing the entire study to a close became pressing, and with 484 questionnaires on hand, representing 95 groups in 19 states, it was felt that the effort should cease.[2]

With two exceptions, wherever there were returns from members of a group a questionnaire filled out by the group leader was also forthcoming. Although some leaders reported that their groups had disbanded, they were nevertheless asked to fill out a leader's questionnaire, both because the group seemed to have been active and relevant to the Study, and because it was desired to have a record of a few groups that had failed. A few other leaders were asked to complete questionnaires, but were not asked to distribute questionnaires among their members, as the time remaining was too short to permit this. Several others sent in their own questionnaires, but reported that their members did not wish to coöperate. Thus 105 questionnaires from leaders were received and tabulated, representing that number of groups, while, as stated, members' questionnaires from only 95 groups were received. However, the groups still active, as represented either by members' or leaders' questionnaires, or by both, totaled 102.

It will be useful for the careful reader of the following pages to know the programs which are represented by the questionnaires of all kinds, the number of groups that listened to each program, the full number of members that heard each program, and whether or not leaders' questionnaires were received in each case. Two tables, one dealing with listening groups and the other

---

[2] Altogether, 446 registered listeners and 521 group members returned their questionnaires. However, a few of these were rejected as too fragmentary in character—5 registered listeners' and 16 members' questionnaires. Others came in after tabulation was begun. Therefore but 423 registered listeners and 484 members are represented in the statistical totals. Late questionnaires raised the total number of groups represented by the members' returns to 97, and the number of states from which members reported to 22.

with groups of registered listeners, have been prepared to cover such information, and are presented here as Tables I and II.

TABLE I

PROGRAMS REPORTED BY GROUP MEMBERS AND LEADERS, AND THE NUMBER
OF GROUPS AND MEMBERS HEARING EACH PROGRAM

	Title of Program	Number of Groups	Number of Leaders Making Reports	Total Number of Members
1	Agricultural Program (Ky.)	4	3	18
2	Agricultural Program (Louisiana State Univ.)	2	2	9
3	America's Town Meeting of the Air	27	30	198
4	Arizona Gardens	—	1	—
5	Democracy in Action	1	1	6
6	Development of the Child (Michigan P.T.A.)	1	1	2
7	Education, Please	2	2	7
8	English, French Programs (Ky.)	—	1	—
9	Family Life Radio Forum (Okla.)	10	10	42
10	Family Life Series (Calif.)	19	22	77
11	French Program (Louisiana State Univ.)	3	3	17
12	Great Plays	—	1	—
13	Homemakers' Forum (N. J.)	10	10	28
14	"It's Your Future" (Iowa)	1	1	14
15	Kay Kyser's College of Musical Knowledge	2	2	11
16	Lux Theatre Program	1	—	1
17	Metropolitan Opera	2	2	4
18	NBC Music Appreciation Hour	1	—	3
19	News Programs	—	3	—
20	Open Forum	1	—	10
21	Radio Child Study Club (Iowa)	5	6	27
22	Symphonic Varieties	1	1	5
23	West Virginia P.T.A.	1	1	2
24	Wisconsin P.T.A.	1	1	3
	Various Programs	—	1	—
	Totals	95	105	484

TABLE II

PROGRAMS REPORTED BY REGISTERED LISTENERS

American Novel	13
America's Town Meeting of the Air	74
Bible Class	4
Columbia's Camera Club	21
Early Risers Club	13
Economics Program	1
Family Life	11
Farm Service	2
French Lesson	243
Greek Epic in English	7
Kay Kyser's College of Musical Knowledge	1
Lady Lend an Ear	15
Our Church on the Air	4
Pulse of Business	5
University Civic Theatre Players	6
Writers' Round Table	3
Total	423

Some readers will also wish to know what radio stations were named as broadcasting programs to the leaders, members, and registered listeners who returned questionnaires to the Study, and in what states the stations were located. Table III presents this information.

Finally, a table has been prepared (Table IV) to show by states the number of leaders, group members, and registered listeners reporting, and also the number of groups represented by members' questionnaires—since these are not identical with the groups for which leaders' reports were received.

As an indication of the full work done in dealing with groups, it may be noted that a total of 410 letters of inquiry were sent out to supposed group leaders or persons who were reported to know about specific groups. As a result, 191 replies were received. Altogether, 138 specific groups were heard from, and 116 of these promised to coöperate by filling out questionnaires. In the end, 122 did so, although 19 of these sent in completed questionnaires for leaders only.

TABLE III

RADIO STATIONS, LISTED BY STATE AND CITY, FROM WHICH PROGRAMS WERE
HEARD BY PERSONS FILLING OUT QUESTIONNAIRES FOR THE STUDY *

Station	State	City	Groups Hearing Station, as Reported by Leaders	Groups Hearing Station, as Reported by Members	Registered Listeners Reporting Station
KTAR	Arizona	Phoenix	1	—	—
KVOA		Tucson	1	—	—
KECA	California	Los Angeles	—	—	6
KFI		" "	—	—	1
KHJ		" "	12	10	—
KNX		" "	—	—	16
KFSD		San Diego	1	1	1
KFRC		San Francisco	7	6	—
KGO		" "	1	2	9
KSAN		" "	1	1	—
KLZ	Colorado	Denver	—	—	44
KOA		"	—	—	1
WICC	Connecticut	Bridgeport	2	2	—
WGAU	Georgia	Athens	—	—	4
WSB		Atlanta	3	3	—
WBBM	Illinois	Chicago	1	—	1
WENR		"	5	4	11
WMAQ		"	—	—	1

* This table shows only those stations reported by persons filling out questionnaires. No listing has been made of the total number of stations reporting listening groups or registered listeners to the Study. Such a list would naturally be much longer. Sometimes a leader, member, or registered listener failed to report the radio station from which he heard a program. Sometimes he named several stations as broadcasting the series to which he listened. Accordingly, the totals given above do not necessarily match those in Table IV for the full numbers of leaders and groups.

TABLE III
(Continued)

Station	State	City	Groups Hearing Station, as Reported by Leaders	Groups Hearing Station, as Reported by Members	Registered Listeners Reporting Station
WBOW	Indiana	Terre Haute	—	—	9
WOI	Iowa	Ames	3	5	1
WMT		Cedar Rapids	—	—	2
KSO		Des Moines	1	1	—
WSUI		Iowa City	2	3	21
KMA		Shenandoah	1	1	—
KGGF	Kansas	Coffeyville	1	1	—
WREN		Lawrence	—	—	1
WLAP	Kentucky	Lexington	4	—	—
WHAS		Louisville	4	4	—
WJBO	Louisiana	Baton Rouge	3	3	—
KVOL		Lafayette	2	2	—
WDSU		New Orleans	—	—	3
WBAL	Maryland	Baltimore	—	—	1
WFBR		"	—	—	1
WBZ-A	Massachusetts	Boston	1	1	4
WEEI		"	1	—	—
WNAC		"	1	1	—
WBCM	Michigan	Bay City	—	—	1
WXYZ		Detroit	1	1	2
WKAR		East Lansing	1	1	—
KSTP	Minnesota	Minneapolis	—	—	1
KFUO	Missouri	St. Louis	—	—	3
KMOX		" "	—	1	—
KGVO	Montana	Missoula	—	—	1
KOIL	Nebraska	Omaha	1	—	—

TABLE III

(*Continued*)

RADIO STATIONS, LISTED BY STATE AND CITY, FROM WHICH PROGRAMS WERE
HEARD BY PERSONS FILLING OUT QUESTIONNAIRES FOR THE STUDY

Station	State	City	Groups Hearing Station, as Reported by Leaders	Groups Hearing Station, as Reported by Members	Registered Listeners Reporting Station
WOR	New Jersey	Newark	8	10	—
WJZ	New York	New York City	8	8	18
WNYC		" " "	1	1	—
WHAM		Rochester	—	—	1
WGY		Schenectady	—	—	1
WSYR		Syracuse	2	2	—
WADC	Ohio	Akron	1	—	—
WLW		Cincinnati	4	1	—
WHK		Cleveland	1	—	1
WTAM		"	1	—	—
WCOL		Columbus	1	1	—
WOSU		"	—	—	243
KADA	Oklahoma	Ada	—	1	—
KCRC		Enid	3	3	5
KBIX		Muskogee	—	1	—
WNAD		Norman	1	—	2
KTOK		Oklahoma City	2	2	1
KGFF		Shawnee	2	2	1
KOIN	Oregon	Portland	—	—	1
WSAN	Pennsylvania	Allentown	1	1	—
KDKA		Pittsburgh	—	—	4
WSM	Tennessee	Nashville	—	1	—
WRTD	Virginia	Richmond	—	—	1
KJR	Washington	Seattle	2	2	1
KGA		Spokane	1	2	1
KMO		Tacoma	1	1	—
WWVA	West Virginia	Wheeling	1	1	—
WHA	Wisconsin	Madison	1	1	—
		Totals	104	95	427

TABLE IV

STATES IN WHICH LEADERS OF LISTENING GROUPS, MEMBERS OF GROUPS, AND
REGISTERED LISTENERS REPORTED THEMSELVES AS BEING LOCATED

State	Number of Leaders	Number of Members	Number of Groups Represented	Number of Registered Listeners
Arizona	1	—	—	—
California	25	100	22	29
Colorado	—	—	—	36
Connecticut	2	8	2	—
Georgia	3	13	3	3
Illinois	6	29	4	21
Indiana	—	—	—	7
Iowa	8	59	10	20
Kansas	1	4	1	—
Kentucky	5	22	5	2
Louisiana	5	26	5	3
Maryland	—	—	—	1
Massachusetts	2	6	2	2
Michigan	2	10	2	10
Minnesota	—	—	—	1
Missouri	—	10	1	6
Montana	—	—	—	1
Nebraska	1	—	—	1
New Hampshire	—	—	—	1
New Jersey	13	40	13	1
New York	8	76	7	6
Ohio	2	3	1	226
Oklahoma	11	42	10	12
Oregon	—	—	—	2
Pennsylvania	2	8	1	11
South Carolina	—	—	—	1
South Dakota	—	—	—	1
Vermont	1	—	—	—
Virginia	—	—	—	2
Washington	4	23	4	2
West Virginia	2	2	1	—
Wisconsin	1	3	1	3
Wyoming	—	—	—	2
Canada				1
No Answer				9
Totals	105	484	95	423

A word remains to be said about visits to groups.

From the reports about listening bodies made to the Study by questionnaire, by letter, and in person, the addresses of many leaders and promoters of groups were obtained. Those in charge of programs with large group followings also furnished the names of leaders, as these were desired. Arrangements were then made by the director to visit certain groups, listen to programs with them, and talk with their leaders. About six weeks were taken up with such field work, and groups in session were observed in six states and the District of Columbia. Altogether, sixteen listening bodies were seen in action, representing various types of persons and covering a variety of programs. A larger number of group leaders and organizers, meetings of whose groups could not be visited, were also interviewed. While an even greater amount of this personal questioning and observation would have been undertaken had time permitted, it was felt that enough had been done to serve as a check upon questionnaires and letters, and to cover most patterns of group action.

Throughout the Study, the director was assisted by advice and suggestions. Levering Tyson, as the representative of the Federal Radio Education Committee, was helpful both in planning what should be done, and in discussing the progress of the work. Leonard Power, coördinator of all the FREC studies, also tendered valuable suggestions, and together with John W. Studebaker, the chairman of the Committee, made possible the sending out by the U. S. Office of Education of the three questionnaires referred to above. Paul F. Lazarsfeld, director of the Office of Radio Research at Columbia University, and formerly in charge of the Princeton Radio Research Project, also gave valuable advice from his extensive experience, as did several members of his staff, particularly Edward A. Suchman. Suggestions were also made by J. Wayne Wrightstone of the New York City school system, who was associated in 1939 with the FREC project for the Evaluation of School Broadcasts, located at Ohio State University. As the Study drew to a close, Dr. Lazarsfeld arranged to have his staff tabulate at cost the returns from the questionnaires for registered listeners and members of listening groups—a highly important task.

Officials of the National Association of Broadcasters, of the National Broadcasting Company, and the Columbia Broadcasting System, particularly Frank M. Stanton, also gave encouragement and specific suggestions. George Denny of Town Hall, and his associates, Chester Snell, Byron Williams, and Arthur Northwood spared no effort in providing the Study with facts which they themselves had gathered from their groups by questionnaire, and in giving lists of individual listeners and the addresses of group leaders.

Numerous officials in charge of radio programs, in addition to other persons previously mentioned, gave the Study aid. I wish particularly to thank Miss Alice Sowers, director of the Oklahoma *Family Life Radio Forum;* Mrs. Marion F. McDowell and Mrs. Phyllis B. Davis of the New Jersey *Homemakers' Forum;* Mrs. Margaret-Elisabeth Tapper and Ralph Ojemann of Iowa's *Radio Child Study Club;* and Mrs. R. W. Marvin, Mrs. V. M. Burrows, and Mrs. L. T. Bravos of the California Congress of Parents and Teachers.

Finally, I am indebted to my assistant, Miss Eleanor Shearer, for her able management of the complicated routines of the Study, for making up numerous statistics, assisting in the tabulation of the returns from the questionnaires for leaders, and helping to prepare the manuscript for the press.

FRANK ERNEST HILL

New York, N. Y.
December, 1940

I · LET'S LISTEN TOGETHER

ONE Saturday afternoon in early October, 1925, the book editor of the New York *Sun* was walking down that twisting street in Haverstraw, New York, which leads to the West Shore railroad station. He was not in a hurry. There was still half an hour before a jerky local would arrive to take him to Weehawken. So he did not mind when he found the sidewalk before him blocked by a knot of men and boys, their tense faces toward the doorway of a store from which came the voice of a radio loudspeaker.

The journalist paused. Play by play, the World's Series struggle between the Yankees and the Giants was being reported. He became one of the sidewalk group. He heard the calling of balls and strikes, and the crack of bat against ball, mushrooming out into the roar of bleacher fans. Soon he decided to wait over until the next train. A little later, he let that go by, too; anyhow, the next one was an express. He stood on the sidewalk listening until the game ended in cheers.

At this time there was no Federal authority regulating radio activity in America, although about three and a half million Americans possessed radio sets. When the air waves could bring men and women into immediate contact with happenings far away, people gathered about the nearest loudspeaker they could find. For an hour or a fraction of an hour they made a group, but it was assumed they listened together because it was the only way in which they could listen.

Such gatherings were of course not listening groups as we now think of them. Today, the term refers only to groups which meet regularly, with the idea of carrying on a continuing activity. Such groups go back far in the brief history of radio. Ernest La

Prade of the National Broadcasting Company, a veteran in radio, told the writer recently that he could not remember the time when groups did not exist. We have positive evidence of their being active in America as early as 1928 or 1929, when parents were listening together to broadcasts from Station WEAO at Ohio State University. In 1931 the National League of Women Voters organized bodies of listeners on a national scale; during that year 20 units in 9 different states were functioning.[1] British activity of a similar kind existed as early or earlier.

In fact, the English seem to have led other European nations and the United States in the promotion of group listening. Educational officials of the British Broadcasting Corporation discussed the question of groups as early as 1927, and seem to have begun organizing them in 1929. A later commentator records fifty groups as having functioned during that year.[2] Swedish, Czech, and Norwegian activity did not develop until the 1930's, and German groups do not appear to have been formed until the turn of the decade. Meanwhile, R. L. Lambert of the British Broadcasting Corporation reported in 1930 to the Institute for Education by Radio at Columbus, Ohio, that "we have a movement to promote discussion or study groups among listeners," and stated that there were then "about two or three hundred such groups of adults, with an average attendance of twenty-five to thirty in each group." But he warned his American listeners that such units were a matter of "slow experiment."[3]

Thus the first British groups had the encouragement and guidance of national radio authorities, even if the official helping hand was a somewhat hesitant one. American activity was in contrast haphazard. It leaned upon the enthusiasm of individuals, voluntary organizations, or institutions. Nevertheless, American group listening from 1931 onward showed a remarkable vigor. In 1932 the two educational radio stations, WOI and WSUI, representing respectively the State College and the State Uni-

[1] *Education on the Air, 1930* (Ohio State University), pp. 168 ff.; *Education on the Air and Radio and Education, 1935* (Ohio State University), p. 171.
[2] *Group Listening*, Bulletin No. 8 in Information Series of National Advisory Council on Radio in Education (University of Chicago Press, 1934), p. 28.
[3] *Education on the Air, 1930* (Ohio State University), p. 104.

versity of Iowa, successfully promoted groups for a child welfare program which is still broadcast today as the *Radio Child Study Club*. In the same year many groups sprang up voluntarily—at colleges, Y.M.C.A.'s, and clubs—to listen to some of the series of broadcasts sponsored by the National Advisory Council on Radio in Education. Early in 1932 another program especially designed for group listening was launched from Station WOR by the Extension Service of Rutgers University, the New Jersey College of Agriculture, and the New Jersey Congress of Parents and Teachers. In that first year, according to "literature" of the Extension Service, 161 groups and 1,640 individuals enrolled for the talks which marked the beginning of a long-continued activity, still going forward under the program title of *Homemakers' Forum*. In 1933 the University of Kentucky began to establish its well-known listening centers, which developed their own types of groups.

Why were the first groups organized? Formerly their existence was attributed to a scarcity of receiving sets. I was told in 1937, when making an extensive survey of education by radio, that groups had sprung up chiefly for this reason; and even in 1939 I was assured by local radio officials in several cities that their communities had no groups because practically every family had at least one radio!

A scarcity of loudspeakers undoubtedly helped for a time to promote many European and American groups, but it is quite clear that this scarcity was not a fundamental factor in group development. Even in 1930 educators perceived that group listening had advantages and attractions of its own with which individual listening could not compete. Also, while the number of American receiving sets has grown to a total of 44,000,000, the groups—as we shall see later—have also increased in number. Clearly the impulse to listen together exists regardless of the distribution of receivers.

Indeed, it can be observed on every hand. "Let's listen to Town Meeting," says one member of the family, and the family gathers to hear the program. "Come over and listen to the opera," says one neighbor to another. "The first program in this series is going on in a few minutes," remarks a radio executive to one of his associates. "Let's hear it and compare notes." Fan mail to almost any radio program, and especially to those carrying information, will include such comments as, "Our family always listens together to your program," or, "Some friends of ours listen with us regularly to your broadcast." Few Americans, if they catechize themselves severely, will deny that on numerous occasions they have wanted others to listen with them, and would have been disappointed, or have been disappointed, if fellow listeners could not be found.

This desire for companionship in listening is probably nothing more than a common human impulse to share experience. Why do men and women want to go to the theater or to a concert together? Why do most of them dislike drinking alone? Why do they like to converse with each other, or read aloud? Obviously the human animal wishes to have experience with his fellows, to talk with them about what has been experienced, if only a few words are spoken. And if he is to engage in more serious activity, such as some purposeful self-improvement, often he wishes to undertake this in common with others inspired by similar motives.

There are of course other factors which make for regular listening together. One of these continues in a fashion the influence once exerted by a scarcity of receiving sets. I may have a radio set in my own home, but I am, let us say, a young man interested in serious music. I have younger brothers and sisters, and they do not share this interest. On Saturday afternoons, or on Saturday evenings, or Sunday afternoons, they are about the house, noisily playing, or wanting to hear a broadcast of a baseball game or a hockey match or a quiz program. But the local museum provides a room where music can be heard. The surroundings are peaceful and attractive. I join a group of others in the museum—people who may talk about the program after it has been heard, and will certainly be quiet while it is in progress. Or I go to the house of a friend who has a better radio than mine and also a family devoted to music. The same thing occurs, perhaps, with programs like the *People's Platform* or *The World Is Yours,* or a program about photography, which—"I" am now a different person—is a hobby of mine.

Or it may be that I am a parent with children, and want to know, as fully and accurately as I can, how to deal with them. I am willing to study child psychology seriously, but I know I am not likely to do so if I depend upon my own lonely efforts. I learn that a group meets at the house of a neighbor to hear a series of radio talks or conversations by experts in child welfare. This group receives printed materials to assist in a better understanding of the problems discussed. I know that if I join this group I will listen attentively to the broadcasts, whereas if I remain at home I may be interrupted while listening, or even forget to tune in. I know there will be discussion which I may find revealing, and I may be delegated to do certain reading, which I have time to do but would not do except in connection with the group. I join these listeners in order to be sure to hear programs that may help me, and to fortify my serious impulses by informing discussion and work.

Thus a common impulse to listen together is shared by many Americans. Some do little with the impulse. Others are led by it into a more intensive experience which may, and as we shall see often does, alter their habits and actions, sometimes profoundly. This impulse lies back of all radio listening group activity. There is no doubt now that it is little affected by the presence or absence of a receiving set in an individual's home. Let us see what it has done to Americans. We shall find that what has already happened as a result of this impulse is startling, and that the possibilities for happenings in the future are even more significant.

II ▾ SPECIMENS

WHAT in Hades is a radio listening group?" impatiently (and a shade more profanely) asked a correspondent to whom I had written a personal letter on the stationery of the Study as a result of which this volume is being written. It is a question still often asked, and one involving complications which even most persons familiar with the activity of listening groups are unlikely to suspect. It is a question necessary to consider at this stage in our exploration.

Let us look at the chief kinds of radio groups which have sprung into existence during the last seventeen years. They are almost as varied in form as the programs which a listener can tap on the dial of a set.

There are what might be called occasional groups; they assemble perhaps only once, or on infrequent occasions. Some are wholly ephemeral and spontaneous; many such groups assembled late in 1936 when King Edward VIII of Great Britain renounced his crown for love. I was a member of one at the first National Conference on Educational Broadcasting in Washington, D. C. The fact of the King's impending broadcast became known, a receiver and a room were supplied by the hotel in which the Conference was meeting, and in the late afternoon several hundred persons came together to hear this strange modern abdication, appropriately announced by a mechanism which a century ago would have frightened almost any monarch from his throne!

Colleges, clubs, and fraternal orders often convene in a somewhat more formal fashion to hear a message or talk by the President of the United States, or the broadcast of an international club president, or the proceedings at a national convention. Again, a panel of listeners may be gathered to hear a particular broadcast and discuss it. Such panels sometimes function at the request of a radio station or a network which wishes to get the constructive critical crossfire of nonradio opinion. Some presidents of colleges report that they call meetings of their student bodies to listen to four or five significant political or other broadcasts in the course of the school year. A national organization which is sponsoring a new radio series on a network will often suggest to its local units that these become listening groups for at least the first few broadcasts.

We shall not deal in detail with such "listening groups," but it is important to recognize that they exist. They represent the impulse to share experience of which I have already spoken, and some of them involve an attempt to be instructed, or to evaluate a radio offering. In addition, they may lead to the formation of groups which meet regularly.

It is the latter with which we are concerned, and they in turn are of numerous types. I shall not attempt an intensive classification, but it will be well to distinguish now between informal and formal groups.

The first may outnumber the second, and their significance is considerable, not only because of what they do themselves, which may have its own importance, but because they frequently lead to something carefully organized and thorough.

We Listen as a Family

In the informal class, family groups are perhaps the most numerous. I remember being invited by the Federal Radio Project to go through a selection of letters from its "fan mail" for *Americans All, Immigrants All.* I found dozens of letters which reported in effect: "We always listen as a family to your program." In conversations with those in charge of radio programs for various organizations or stations I found few persons who did not testify to receiving letters from such "groups." So far as I know, there has never been an attempt by any program representative to make a list of them or to establish more than fleeting contacts with their members. Yet they must number tens of thousands. Naturally, some are likely to be irregular, to dissolve (perhaps tiring of a particular program and turning to another), or after a time wholly to cease group activity. Still, I do not doubt that many have persisted for long periods, have held their informal discussions, and have stimulated their members to read, to visit museums and galleries, or to take up hobbies.

Informal groups of several families, or of various friends, also exist. I have stumbled on them here and there as I visited organized listening groups or talked with persons interested in radio work. Others have been reported to me. An investigator in almost any progressive community, interviewing one hundred Americans as to their radio activities, would be likely to turn up at least several such units. They may gradually take on the character of organized groups.

"My husband and I never made an engagement on Thursday evenings outside of our house," writes a woman in Massachusetts to Town Meeting headquarters. "Our discussions often became so heated that it was two or three o'clock before we could calm down enough to think of sleep! This year we decided to ask ten of our friends to meet with us. The group became so interested we decided to make a supper club of it, in order to give more time for discussion." This particular group now meets at a tea house, and has a membership of fourteen men and sixteen women. Many well-organized listening groups have developed from casual beginnings of a similar kind, gradually acquiring form, purpose, and a larger membership.

Other types of informal groups have come into being in the radio rooms established by museums, libraries, colleges, CCC camps, NYA and WPA centers, Y.M.C.A.'s, community houses, clubs, and even radio stations! Usually certain programs attract a core of regular listeners to such rooms, although the larger part of the audience may change from day to day. Whatever discussion takes place is usually spontaneous and unplanned. Yet such groups are notable for their number and their attentive interest, and are often the beginnings of a consciously regulated activity.

The formal, organized groups naturally attract more attention and usually accomplish more than those described above, which represent a continuing yet rather casual interest. The formal groups are our particular concern, so let us examine them rather carefully.

Species and Subspecies

At first glance most of them seem much alike. They are bodies whose members gather at regular intervals to hear a broadcast; they have a leader and sometimes other officers; they follow specific procedures as to listening and as to activities supplementary to listening. Such, at least, is the usual pattern. Some units lack certain of these characteristics, and there are almost infinite variations as to the way in which discussion, study and social activity are carried on from group to group.

Perhaps purpose and subject matter are most likely to make for variety. As to purpose, the following types exist:

Groups which listen to a broadcast merely for general information, and discuss it only casually.

Groups which listen in order to use the broadcast as a springboard for organized discussion.

Groups which listen chiefly to get stimulus and suggestions for the study of a particular subject.

Groups which have both discussion and study as their objectives.

The subject of the broadcasts naturally has an influence on the character of the group. Two types of subjects have attracted rather sharply defined groups: public affairs and family life problems. The first is well represented, as to programs, by *America's Town Meeting of the Air,* the *Chicago Round Table,* and the *People's Platform;* the second by the New Jersey State Agricultural College's program, *Homemakers' Forum,* or by the two Iowa universities' program, *Radio Child Study Club,* or by the *Family Life Radio Forum,* broadcast from the University of Oklahoma, at Norman, Oklahoma.

More groups listen to these two general classes of programs than to all the other types combined, for the obvious reason that the political and social progress of the nation and the welfare of the family are, respectively, subjects which arouse a keen interest in the majority of Americans. Doubtless the importance of the two general subjects has helped to encourage the production of excellent programs and of services to groups—both in turn definite stimulants of group activity.

Nevertheless, there are other subjects which attract group listeners. Music is one. In the past, two programs which encouraged the playing of music by listeners won a notable popularity. Both attracted numbers of groups.

Dr. Joseph Maddy, originator of *Maddy's Band Lessons*, had 10,000 individuals or groups using his study aids for a time, and, over a period of years, more than 250,000 scores for individual instruments were sold to listeners to the *NBC Home Symphony*, many of these going to groups. Listeners to both programs played with the band or orchestra used as features of the broadcasts. However, today Dr. Maddy's program is no longer a national network feature, as it was for several years, and *NBC Home Symphony* has "left the air."

During the period when his broadcasts went over the network, Dr. Maddy traveled constantly, visiting groups and often using them in his broadcasts. Most of his groups were in schools, a fact which makes them of less interest to us, since, as stated in the Foreword, we are primarily concerned with adult groups, including those composed partly or wholly of college students. The musical program sponsored by the Standard Oil Company of California has also been used mostly by schools, but if we are to count classes as listening groups, the more than 3,000 schools and 325,000 children which it serves represent a tremendous and carefully organized activity.

Other musical programs attracting group listeners are the Metropolitan Opera broadcasts, those of the NBC Symphony Orchestra, and the Philharmonic broadcasts.

A miscellany of other programs attract organized groups. *Great Plays* claims on good evidence to have more than 1,000; news broadcasts are used by a number of groups; garden clubs listen to the Mutual Network's *Radio Garden Club;* agricultural programs of various types are heard by groups; and *Education, Please, Democracy in Action, On Your Job, Art for Your Sake* and other series have their regular bodies of listeners. The series, *You and Your Government,* which ran from 1932 until 1936, called into being many groups, as did the *New Poetry Hour,* broadcast over WOR and the Mutual Network, and other programs no longer on the air.

Groups of Registered Listeners

One type of listening group, different from any of the above, informal or formal, should be described here. Its peculiar distinction is that its members listen alone. Yet they are certainly a group. Station WOSU at Ohio State University, for example, broadcasts French lessons. During the year 1938-39, a total of 798 persons wrote to the station to get the outlines essential to an effective use of the course. At any broadcast a number of thousands, perhaps tens of thousands of listeners, were undoubtedly before the loudspeaker; among these the 798 made a special group. I have called such bodies of persons groups of registered listeners, since each one of them, by putting himself on a list to receive service, "registers" with the station or program headquarters which serves him. The whole number of registerees constitutes the group.

There are groups of registered listeners in most states of the union. Responses mailed to the Study from 423 such persons came from 28 states and the Dominion of Canada. Registered listeners hear programs on art, on marriage, on American literature, the Greek masterpieces, the mechanics of radio, photography, various phases of religion, music, domestic science, and many other topics. A number of such groups take calisthenics every morning in various parts of the United States. A group of individuals who pay for study aids listens to *America's Town Meeting of the Air*—254 did so in 1938-39—and of course this "group" is wholly different from the regular listening groups for Town Meeting, which meet together at one place, while the registered listeners listen in 254 different places, chiefly as individuals.

It would have been pleasant to study all types of radio listening groups, from families and panels and clubs undertaking only occasional assignments to classes in schools and organizations with elaborate speaking programs. I have not been able to study them all, and can deal only with the better organized types apart from the grammar and high schools. These are numerous enough. They comprise all the formal adult and college groups described above, and the registered listeners.

We shall give them full consideration. We shall walk into their meetings, talk with their leaders, get their written depositions about their work and their desires, examine the arrangements which radio stations and program directors have developed for serving groups, and search weaknesses and strengths along the way. We shall even find out how to call a group into being, and how to make it prosper. And we shall ask what effect groups have, through their activities, upon radio stations, upon themselves, and upon American life in general. If the groups do not like this fairly comprehensive if tentative probing into their affairs, they have themselves to thank. They have flourished and become important. Educationally and socially, they represent a new national habit. They will perhaps change America in important ways, and America needs to understand them.

III · UNFINISHED CENSUS

NOBODY knows how many radio listening groups there are in the United States, and it is improbable that anyone ever will know. Born of the impulse to listen together, some of these find a firm and persisting life, others quickly dissolve. Mortality and new birth represent constantly varying factors, and today's census, could it be taken, would not be tomorrow's.

Nevertheless, it is possible to get some idea as to the prevalence of groups in America, and it is highly desirable to have such an idea. In making a study of groups, I asked constantly for reports on their number, and got definite testimony from various agencies which dealt with groups and knew about them. The sampling process by which this incomplete census was taken has been described in the Foreword; it comprised questionnaires and inquiries directed to radio stations, colleges, libraries, CCC camps, NYA state directors, labor councils, and directors of radio programs. The results, listed here according to the sources from which they came, are as follows:

Reported by	*Number of Groups*
Radio stations and networks	6,507
Colleges	565
Civilian Conservation Corps	583
Labor councils	105
Libraries	47
National Youth Administration (estimated)	50
Family Life Radio Forum (Oklahoma)	242
Homemakers' Forum (New Jersey)	214
Metropolitan Opera Guild (estimated)	50
Total	8,363

Skeptical Tabulation

These totals stand for groups definitely reported by the agencies concerned. For those who wonder exactly what is meant by "definitely reported," I shall explain the phrase.

Let me say first that although any investigator interested in the extent and forms of a particular activity dislikes to find that he has nothing to investigate, I was aware from the first of dangers of exaggeration in connection with reporting groups. I understood clearly that my task was not to make a case for the existence of groups. If there were no groups, that would be a fact quite as important to me as the possible fact that there were many. Similarly, if the activities of most groups were perfunctory and of little significance, it was clearly my business to report the disappointing character of their work. I have consequently tried throughout the study to detect and discount exaggeration. Therefore I held closely to an attitude of healthy skepticism in compiling the list given above.

Many stations listed radio programs which had listening groups, but unless a specific number of groups was given, or some such phrase as "at least one," was used, no groups were counted. If "at least one" or "at least two" was set opposite the program title, credit was given for one or for two groups, according to the number stated. Even if the station reported "number unknown," indicating that groups did exist, no credit was given. Also, if it seemed uncertain as to whether numbers of groups or numbers of persons in one group was meant, only one group was recorded. For example, a Virginia station reported 318 groups for a college program, but this report was accompanied by no explanation and no addresses of leaders, and credit for a single group was given.

Even network reports for programs like the *NBC Music Appreciation Hour* and the General Federation of Women's Clubs' program, *Adjusting Democracy to Human Welfare,* when accompanied by the comment, "number of groups unknown," were accepted only as "leads," and no groups tabulated. Undoubtedly both these programs had groups. In the second case an inquiry directed to the Federation brought the statement that no reports from local clubs on group listening had been made to the national organization, so that there was a double negative reason for giving no credit, although the implication was that a number of listening units had existed. Again, Station WCCO in Minneapolis reported that fifty-six local organizations had agreed to listen to the program, but that it was unknown how many had done so. No groups were credited to WCCO.

In no instance were "leads" recorded as groups, even when the names and addresses of probable group representatives were given. Only a clear declaration by the station or network that a certain number of groups existed was accepted as authentic.

Consequently, the list given above represents only such groups as the stations or other agencies positively asserted to exist. There was no specific reason to believe groups so enumerated were the result of wishful thinking on the part of station managers, college presidents, labor councils, or the directors of programs. However, because in a few cases such thinking was suspected, and because a check seemed desirable in the general interest of accuracy, an effort was made to verify independently a number of the figures presented.

Verifying the Totals

There was opportunity for verification. For example, in the case of three family life programs I was permitted to examine the reports made by individual groups, to correspond with and visit group leaders, and to attend meetings of groups. As a result of such experience, I have confidence in the authenticity of the figures presented by the respective directors of the *Family Life Radio Forum*, broadcast from the University of Oklahoma's station, and of the *Homemakers' Forum*, directed by the Home Economics officials of the New Jersey State College of Agriculture. Similarly, I saw evidence of the existence of the 175 groups (the total is now larger) reported by Station WOI at Ames, Iowa, for the *Radio Child Study Club*. Thus 631 groups on the list may be regarded as having been in existence during the season 1938-39.

Another large cluster of groups reported was the 1,631 units listening to the broadcasts of the California Congress of Parents and Teachers. As a result of correspondence with the officials of this organization, and of personal reports made to me by radio council chairmen and actual leaders of numerous groups in relatively small areas, I see no reason to doubt the validity of this California report. I had extensive evidence that the bodies listening to the California *Family Life Series* were both numerous and active.

Another body of groups which it was possible to check was that of *America's Town Meeting of the Air*. The officials of the National Broadcasting Company reported a total for these of 3,700. I was skeptical with regard to this number, and took up the question of its validity with the officials of Town Hall, Inc. They showed me evidence of more than 1,900 groups in touch with their Advisory Service, including more than 1,000 CCC, WPA, and NYA bodies. They explained, and showed me evidence to this effect, that classes in entire school systems also listened as groups, as did a number of college classes. In their opinion at least 5,000 groups listened, since they were continually hearing (as I had heard myself) of groups that did not pay for or receive regular service, although their own lists contained only groups which either paid for this service, or were given it because of relief status. We finally agreed upon 3,000 as a figure which would certainly be less than the actual number. This figure is included in the totals for radio stations given above.

A total of 1,000 groups listening to broadcasts of *Great Plays* was accepted after a conference with a representative of that program. Perhaps from 1,200 to 1,500 of the groups for these two programs are composed of students. However, such students do not listen as classes, although many discuss one or the other of these programs regularly in connection with class work. Of course, if classes hearing broadcasts during school hours were to be counted here as groups, we could more than double, perhaps multiply by five, the list given above by adding the "groups" listening to the C.B.S.'s *American School of the Air*, the *NBC Music Appreciation Hour*, and other programs devised specially for schools. As previously stated, no study of classroom broadcasting is included in this volume, and only a few classes are listed as groups.

Altogether, then, some check has been made upon 6,262 of the 8,363 groups listed above. There seems no reason to doubt the validity of the 565 groups reported by colleges, the 105 reported by labor councils, or the 47 reported by libraries. The CCC groups were reported by camp advisers, and I made no check upon these except by correspondence with a few group leaders who filled out questionnaires for me. The NYA and Metropolitan Opera groups represent my own estimates, and I think very conservative ones, made on the basis of correspondence and personal discussion.

So far as P.T.A. and Town Meeting programs were concerned, all duplications (e.g., a report by a station of an *America's Town Meeting of the Air* or a *Homemakers' Forum* group) were eliminated. This was not done in the case of CCC or college or library returns, and perhaps from 150 to 300 groups in these returns represent duplications. However, these are more than offset by the several hundred miscellaneous groups located by the Study in following up "leads." These are not included in the table given above. Everything considered, including a probable effort by some of those reporting to make the best showing possible, it seems certain that as many as 8,363 groups definitely functioned at some time during the season 1938-39. Many of those reporting groups gave the number of group members. For the others, an average of 10 per group was arbitrarily taken as a probable number of members—certainly an underestimate.[1] The resultant total membership of all the groups was 155,840. Personally, I do not doubt that it was more than 200,000, as 15 or 20 would be a more likely average for group membership than 10.

How Many Groups in America?

Remembering that I was frequently asked by educators and radio officials who were told that I was making a study of radio listening groups: "Will you find any?" I contemplate the above figures with some astonishment, although I was not unprepared

[1] The average membership of 105 groups for which questionnaires were received by the Study was 56.2 persons. Omitting, as exceptional, two groups that reported 1,000 and 2,000 members, respectively, the average for the remaining 103 was 29. P.T.A. groups tend to be small; Town Meeting and musical groups large.

to find evidence of more than 5,000 groups. Nevertheless, it should be borne in mind that the number of groups listed above can certainly be considered far smaller than the number of organized groups actually functioning.

Probably they included the majority of strong and active listening bodies, although even that is uncertain. No follow-up was made in the case of questionnaires sent to radio stations, colleges, libraries, and other organizations. No account was taken of the carelessness or ignorance of the officials who replied, although later evidence proved that groups had not been reported in many cases because of carelessness or ignorance. In addition, many organizations, like the General Federation of Women's Clubs, the National Education Association, and the Metropolitan Opera Guild, have made no effort to get reports on possible groups. In the course of the Study, only a limited number of persons in charge of radio programs were asked if they had records of groups, and most of them, including representatives of the *People's Platform*, the *Chicago Round Table*, the *American Forum of the Air*, the *Radio Garden Club*, and the *National Farm and Home Hour*, replied that they had kept no record of listeners meeting together. Yet a number of groups were encountered that listened to such programs, and doubtless many more exist which were not encountered because of the limited scope of the Study.

Again, no canvass was made of news programs, quiz programs (with two exceptions soon to be noted), or other semieducational series of broadcasts, the groups engaged in definitely educational activities being considered sufficient to occupy the energies of the Study. Two inquiries, made respectively of the directors of *Kay Kyser's College of Musical Knowledge* and *Information Please* brought replies that undoubtedly there were groups which heard these programs, although no record was kept of them. The Kay Kyser officials, after watching their mail for a week, forwarded the addresses of four group leaders, and two of these responded to letters, and filled out questionnaires. One of these groups contained sixty persons. What might result if groups were requested by radio announcement to report their existence to the dozens of educational and semieducational programs which have hitherto paid no attention to them is interesting to speculate upon.

Furthermore, not thirty percent of the "leads" furnished by agencies reporting on groups were followed up. For example, contacts with P.T.A. groups were established in but seven states, although most state congresses of parents and teachers have listening group activity of some kind. P.T.A. groups possibly number several thousand more than those tabulated above, although the 2,386 listed probably stand for the best organized and publicized listening done by American mothers. Although letters sent to a number of county agents and to a few leaders of 4-H clubs brought in no evidence of definite groups, I am convinced that a considerable number of rural units listen to farm broadcasts, and could with patience and ingenuity be "smoked out." We lacked the time and money to follow them up effectively, feeling obliged to spend our energies on the numerous other groups that could be reached with relative ease.

On the basis of the facts collected, and the possibilities for getting information left ungathered, it therefore seems reasonable to assume that there are in the United States at least 15,000 organized groups meeting together to hear radio programs, and that their activities touch from 300,000 to 500,000 Americans.

Were we to include family groups, friends meeting together regularly but without formal organization, and people assembling in the radio listening rooms of museums, libraries, colleges, and broadcasting stations with some regularity, the total of all group listeners would be greatly increased—perhaps doubled.

None of the listening bodies discussed or listed above include registered listeners. Groups composed of such persons were reported only by radio stations. Altogether, 76 of them were listed, with a total membership of 16,872. Since the radio station is usually the headquarters for such groups, or knows the program directors who deal with them, this total is perhaps a relatively full one, although I have learned of a number of programs having registered listeners which were not reported by the stations where the broadcasts originated.

Obviously, and for reasons which will be set forth later, groups and group members are likely to grow in number. Thus far, little attempt has been made to encourage them; yet, as we shall see, they represent a potential asset for educators and radio stations; and probably they will receive far more attention in the future than they have in the past. Meanwhile, their present numbers give them a definite social significance. They are far more numerous than teachers and radio officials have thought, and by the nature of their activity they are important in the mental, aesthetic and social life of the nation. We can get a more definite idea of their functions and possibilities as we turn to examine them in detail.

IV ⟑ A GROUP IS BORN

I HAVE SAID that there is an impulse in most of us which bids us listen together. Not all Americans are aware of its existence. In fact, whole communities sit contentedly, each family or individual with a radio set, and are undisturbed by the fact that they are missing the satisfactions of organized group listening. How are these satisfactions brought to their notice? How is a group born?

The process takes various forms. It might be said of groups and birth, to paraphrase what has been said of men and greatness, that some are born spontaneously, some struggle hard to achieve birth, and some have it thrust upon them. There is available testimony as to all these processes. Let us look at some of it.

But, first, a word as to how this testimony was gathered. As I explained in the Foreword, one of the purposes of the Study from which this book has grown was to know groups by direct contact, and get evidence at first hand from leaders and members. I have already told in some detail how this objective was carried out—partly by visits to groups, and partly by questionnaires. When group leaders or members are quoted in the following pages, they have spoken personally with the writer; other citations represent the words of some of the 105 leaders, 521 members, and 446 registered listeners, each of whom has answered from 45 to 59 questions about his attitudes, hopes, and activities.

Of course all speculation on why listening groups exist must start with a recognition of the character of broadcasting itself. I shall do no more than refer to this, assuming it to be recognized; yet the potentialities and accomplishments of radio-transmitted programs as these affect education should be borne in mind throughout the present chapter. Broadcasting can invade the smallest towns; it may enter from 30,000,000 to 40,000,000 American homes. Its direct cost is little, or—as to education, since most Americans have their sets in any case—nothing. It permits tremendous audiences, it can offer distinguished speakers; it can give explicit directions to thousands of groups at once; it can be the starting point for the distribution of reading materials or other listener aids. Let us remember this as we consider how groups come to life.

A Few Birth Records

One leader records what may be considered a spontaneous birth. "Dr. Reinhold Niebuhr of Union Seminary," writes a New Jersey clergyman of the occasion which led to the founding of a group, "was a speaker on the Town Hall program. We invited some friends—and they insisted on meeting thereafter."

Sometimes the spontaneity of beginning is due to the fact that a group of persons is already busy with some activity, or has been busy, and turns to radio as a substitute or for an added stimulus.

In Brightwaters, Long Island, a WPA class had drawn together a number of people interested in public speaking. The WPA educational work was discontinued in that community, but one of the members suggested that the speech class continue its work by using the Town Meeting program as a basis for its discussions. This was promptly done, and today the group meets from one to three times a month, and attracts men and women from the breadth of Long Island.

In Plainfield, New Jersey, a number of residents felt the lack of an active intellectual life such as was possible for men and women residing in Manhattan. Particularly they mourned the inconvenience of seeing good plays. This suggested a reading group to Carl Fuerer, a young American citizen of German birth who fought for Germany in the World War. Was it difficult to see plays? "Then we are going to read them!" he exclaimed enthusiastically.

They tried to, but not with full success. Finally the idea came to Fuerer that he and his friends might broaden their experience and interests by listening to *America's Town Meeting of the Air,* and from experimentation with this notion a vigorous discussion club quickly developed.

With the idea that a number of groups might be listening to radio programs because they were already in existence for a purpose which group listening might further, leaders were asked in their questionnaire, "Are listening to the program and talking about it the only activities of the group?" Sixty-nine of 105 leaders replied that they were; 35 that they were not; while 2 did not reply. Leaders were asked to name other activities where these existed, and many types of undertakings were mentioned. These included sewing, dances, lunches, P.T.A. work, the putting on of radio programs, social community activities, and study programs which apparently were sometimes separate from the radio listening and sometimes associated with it. On the whole, few groups seemed to exist chiefly for other purposes, although these purposes appeared to act in a number of cases to maintain the vitality of groups. Their influence is clearly related in most instances to the attitude or actions of sponsoring organizations, which will soon be considered.

Convenience has its part in group birth. In Des Moines, the home of public forums, people have the discussion habit. They also have one night a week when it is convenient for many married couples to dine away from home. "Thursday is maid's day out. Thursday is Town Meeting of the Air," writes the city librarian, Forrest B. Spaulding, in answer to the question of how his listening group came to be organized. Mr. Spaulding and his associates meet for a buffet supper in a residence of one of the group, and then listen to the program at 8:30.

In the case of the Metropolitan Opera broadcasts, a woman near Seattle, Washington, invited a group of friends to hear the opera and lunch with her, and, as she remarks, "we liked it so much that we formed a 'club.'" This group, launched in so casual a way, has a history of eight years' listening.

Struggles, Failures, Mirages

Not always is it so easy to bring a group into being. Often a lone enthusiast labors long and earnestly to collect a few kindred souls, and perhaps never succeeds in collecting them. Gathered for a first meeting, the members may not find the program to their liking, or other activities may attract them more, and the group dwindles away. Guarded phrasing from the letters of supposed leaders who had not been successful suggests how hard is the way of the organizer.

"I tried to organize a Listening Group to listen in on Town Meeting of the Air, but was not successful," writes a "lead" from western New York.

"We do not have any listening group," reports a Maine woman, "as only my sister and myself seem interested in Radio Garden Clubs. Have spoken to several members of the town garden club about the splendid talks we get over WOR, but cannot see that they are at all interested so we just listen ourselves."

"We regret that the organization of our listening groups has not gone forward as we expected," writes a WPA official from Boston. "They are still one of the activities we hope to inaugurate here. If we succeed in getting some of the groups under way, we shall communicate with you."

So the record might be continued. I should like to write at length about groups that have been born by reputation, but never in fact. There are many. The case of Town Meeting groups in St. Louis is a little saga of group mirages, such as anyone who has hunted groups in their native haunts may be able to tell. St. Louis until recently did not receive *America's Town Meeting of the Air*. The hour from 8:30 to 9:30 P.M. is a profitable one for commercial stations, and Station KWK, which had the right to broadcast the program, sold the time. Many citizens bitterly deplored this sacrifice to prosperity. At length the St. Louis Adult Education Council started a campaign to make the broadcasts available to the city, even if only in the form of a transcription played at a less popular hour. Clubs and organizations, urged by the council, signed a petition to put the program on the air, and more than 2,000 signatures were obtained. The station responded, and presented the program at 12:30 on Sundays. Unfortunately the *Chicago Round Table* could then be heard at that hour, and cries of protest were raised. The hour was changed to 2:00 P.M. I arrived at St. Louis soon after the second change had been made, and was told that a number of groups were listening,

several at branch Y.M.C.A.'s and one at the Jewish Community
Center. I promptly called up the central Y.M.C.A.

"Oh, yes," said its representative, over the telephone. "We
have three groups."

He gave me the addresses of the branches at which they met.
I was particularly eager to attend a meeting of one of these
groups, as up to that time I had never met with a group hearing
a program by transcription. I had learned of the groups Friday
afternoon, but being in a hurry at the time, decided to postpone
until the following day the making of my arrangements for
attending. Saturday morning I told an educator of my good for-
tune in finding four groups.

"I'd call up right away and make sure," he advised. "I'm not
certain that those groups are meeting."

I followed his advice. One by one the branches reported that
they had no listening groups.

"We have talked of starting one," said the first official, "but
the time seems a poor one for a group."

"We haven't completed arrangements," said a second, "and
frankly, the prospects aren't very good."

It was the same with the third prospect. The Jewish Community Center also failed to report any group activity, although its
director was much interested in the idea of a group, and plied
me with questions as to how one might be formed and conducted.
Perhaps all these mirage-groups have now become flesh and
blood. When I left the city that Sunday evening they had faded
for the time at least into mere hopes, and St. Louis, which apparently had possessed four bodies of radio listeners a day and a half
before, had none.

Aid from Organizations

Thus it can be seen that groups do not always come easily
to birth. Many workers in adult education who have been responsible for encouraging group listening know this fact only too
well.

"I'd like to get into one room a lot of those people who think
listening groups are easy to start and hold, and talk to them for
a while," remarked Dr. Alice Sowers, the director of the Family

Life Institute at the University of Oklahoma, and the originator of the *Family Life Radio Forum.* "There is literally an endless amount of work to be done if you want to establish and maintain any number of groups."

This assertion brings us to the groups that have birth thrust upon them, and they are numerous. Of the 105 leaders who reported to the Study on their groups, 73 said that their units had been promoted by some organization. In many cases the "promotion" consisted in mere encouragement, or the provision of quarters for meeting, or the payment for study aids; and the initiative for forming and maintaining the group was apparently supplied by individuals. Only 31 leaders felt that their members attended "partly or wholly from a sense of duty." Still, the factor of encouragement by an institution or organization was clearly a strong one.

Beyond doubt the most important aspect of this encouragement is the mere fact that it is given—in whatever form. Many individuals respond to the knowledge that a responsible agency wants to see listening groups function. Some men or women may never have thought of forming a group. The agency brings the idea to their attention, and with the now proverbial tendency of Americans to organize, they leap into action. Particularly is this the case if some assistance is offered to a group or groups even if the latter must pay for it!

The listening groups of *America's Town Meeting of the Air* represent to a large extent the response to the idea of group formation, plus the offer of printed aids to groups. A few listening units already existed when Town Hall, Inc., decided to encourage organized bodies of listeners. Such units may have numbered several hundreds. Yet in less than three years from the first announcements made in the fall of 1937 the total of known groups had grown to 1,960. These were only such bodies as received the regular service sent out by Town Hall.

In the case of the Town Meeting groups, a number, in addition to being serviced by Town Hall, were also encouraged by local organizations such as churches, schools, Y.M.C.A.'s, Jewish community centers, and clubs. These agencies sometimes paid for printed aids, sometimes provided quarters where groups could meet, often furnished leaders, and occasionally rendered all these services. It is difficult to say how much the formation of groups was furthered by the mere endorsement of the idea of radio listening units and how much it was facilitated by the offer of certain benefits such as those mentioned above. Almost always endorsement and definite assistance of some kind were linked together.

Such has often been the case with P.T.A. groups. The seven state congresses of parents and teachers whose groups have reported to the Study all approved programs dealing with family life problems. And all offered one incentive to their locals for the formation of groups: credit, which officially improves the rating of a P.T.A. unit. Many activities are rewarded with credits, but a listening group is a relatively easy instrument with which to lift the standing of a local body. The desire to have the credit which it will bring has undoubtedly given a great impulse to the formation of P.T.A. groups, although, as we shall see, such an impulse is by no means the dominant motive with many of these units.

Thirty-three Kinds of Agencies

Other agencies besides those enumerated above have given encouragement to radio listening groups, some to promote programs of their own, and some to foster activity for Town Meeting or various P.T.A. programs. Among the independents should be listed such universities and states as have originated a series of discussions of family life or child psychology, which P.T.A.'s have used. The Oklahoma, Iowa, and New Jersey programs which have already been mentioned fall into this class. All are developed by their respective universities, and not by the state organizations of the National Congress of Parents and Teachers. In contrast, the California *Family Life Series* is originated by the Congress of that state.

The University of Kentucky has won a just reputation for its work with listening groups in isolated mountain sections. The university, through its radio director, Elmer G. Sulzer, established listening centers as early as 1933, and has had thirty of these, each promoting listening by groups.

Louisiana State University, with the assistance of money supplied by the General Education Board of the Rockefeller Foundation, has also experimented with French-speaking and other groups in Louisiana, establishing a number of listening units (often with the assistance of local educational authorities) and devising and broadcasting programs to which these groups listened.

CCC camps, WPA centers, and NYA centers have also encouraged groups, many of them for programs which gave no assistance to listeners meeting in bodies.

Altogether, among the 73 groups whose leaders reported that their groups were encouraged by institutions or organizations, 33 different types of agencies were reported as promoters. Some groups were assisted by several such agencies—for example, certain P.T.A. groups got listening aids and advice from Dr. Sowers of the Family Life Institute at the University of Oklahoma, and were also encouraged and "rewarded" (with credits for rating) by the state Congress of Parents and Teachers. Again, numerous Town Meeting groups were encouraged and assisted by Town Hall, Inc., and also by churches, Y.M.C.A.'s, colleges, and other organizations.

The following agencies were found participating in listening group activity: Town Hall, Inc.; seven state P.T.A.'s; four universities (through radio officials developing programs); two others by general encouragement; three others through their departments of education, graduate schools, or research projects; Y.M.C.A.'s; Y.W.C.A.'s; Jewish community centers; churches; synagogues; women's clubs; women's university clubs; women's business and professional clubs; high schools; boards of education; libraries; WPA units; CCC camps; NYA centers; national and state garden clubs; and peace councils.

As previously stated, the kind and degree of "encouragement" varied greatly. Altogether, 24 agencies assisted in paying for various kinds of service related to group activity, 22 of them paying all expenses. Other forms of assistance, as already stated, were provision of quarters, in which listening groups could meet, credit for rating, the furnishing of leaders, and the supplying of free study aids.

"We Have To Have Our Rating"

What effect did such encouragement have upon groups? Did they carry on their activities chiefly because of the assistance they received, or from a sense of obligation?

Undoubtedly some did. One P.T.A. group leader who apparently found a minimum of satisfaction in her activities, wrote after the question: "Do members of the group attend wholly or partly from a sense of duty?", "Yes, *wholly!*" At a visit to an Iowa group I tried to probe rather fully into the attitude of the members on the matter of credit for rating.

"Would the group meet if it got no credit?" I asked of four or five members, including a president and a past-president, as we chatted after a really excellent meeting.

"I doubt if we'd meet except for what it means to our rating," answered the president.

"But you are interested in the broadcasts? You feel that they are worth-while?" I persisted.

"Oh, yes, we do. We think they're splendid."

"Well," I remarked, consciously venturing to the verge of insult, "if *I* were really interested in a subject, I wouldn't care particularly what the state P.T.A. thought about my activity."

The ladies exchanged uneasy glances.

"We have to have our rating," said one, firmly.

A day later I encountered an entirely different spirit.

"Oh, we don't care much about the credit," said the president of a group in a neighboring town. "We're glad to have it, of course. But we don't even advertise this as a P.T.A. group. In fact, we keep quiet about the P.T.A. We want to get all the mothers we can, and if some of them thought this was all P.T.A., they'd be afraid we'd ask them to join up and pay dues, and they wouldn't come!"

One gauge of spontaneous interest is the frequency of meetings and the length of the meeting season. Most state P.T.A.'s give credit for listening group activity if as many as 5 meetings are held in any one year. Yet of 50 P.T.A. groups represented by 105 leaders' questionnaires, only 17 held 7 meetings or less in the course of a year, one did not state the number of meetings held, and 32 held more than 7 meetings. Among the latter, 15 met weekly during the "season" from September or October to April, May, or June, 4 met weekly all the year around, and 2 met fortnightly during the entire year. Thus almost two-thirds of these groups met more often than was required for credit, while two-fifths of them held from four to ten times the required number of meetings.

The other "promoted" groups did even better, according to their reports. Nobody suggested to them that five meetings would be enough, and all reported coming together at least monthly, and usually oftener, throughout the "season." However, with them there was no question of rating, and they either continued to meet or languished and died, as two reported doing!

It Did Not Harm Them

On the whole, being "encouraged" did not seem to harm bodies of listeners in the least, and even with P.T.A. groups the tendency was for members, once engaged in listening, to take hold of the job for themselves, and perform it because of interest rather than because of a sense of duty. The role of the agency which stimulated their activity nevertheless remains highly significant. Without the approval and encouragement of such agencies the groups would be far fewer today than they are. To get a strong and respected organization behind a group is one of the best ways of bringing it to life and maintaining its vitality.

In the questionnaire for leaders, two other questions were put which bear on the reason why groups are started. One asked how the group came to be organized. To this, only 39 out of 105 replied that they were asked to organize by some agency, although 73 stated that they were encouraged by such a body. However, 4 reported that they were established as a part of the service offered by some institution or agency, 5 reported that they were classes in schools or Y.M.C.A.'s, 4 were established by the University of Kentucky, and 2 as a convenience or opportunity offered by an organization of which they were members. Thus 54 were essentially the result of organizational effort. Of the remainder, 20 reported that the group was the idea of the leader, 15 that it was established as the result of conversation among friends, while printed announcements were thought to be the chief factor with 2, announcement by air with 2, and a variety of other causes were reported for the remainder.

In another question, leaders were asked to state the purpose of the group in beginning its activity. They could check one of a number of possible purposes as the chief one, and others as contributory. In this tabulation "To provide group discussion" led, with 31 giving it as their chief purpose, and 36 listing it as a subordinate one. "To follow the policy of a national or regional organization" came second, with 28 groups giving it as their chief purpose, and 12 designating it as contributory. "To study" ran third, with 15 and 20; and "To give general information" fourth, with 13 and 42.

Quite clearly the cause of a group's birth varies greatly with the character of the group. The P.T.A. units necessarily show the influence of the national or state congresses of parents and teachers, and they, of course, constituted 50 out of 105 of the groups on which data were available. This proportion is almost certainly a bit higher than the proportion of all P.T.A. groups in the United States to that of all listening groups in the country. Among the 8,363 groups reported to the Study, for instance, they represented less than 40 percent of the total. However, it is probable that half of all American groups have been encouraged or instigated by the varied types of agencies listed above, and in the future the proportion will probably be higher.

The Members Speak

What I have written above is obviously based on the testimony of group leaders, on personal observation, and on interviews with persons having charge of radio programs or of promoting groups. We have an additional source of information: the individual members. What they have to say should be important. For after all, groups are made up of members; and when potential leaders and public-spirited organizations attempt to establish listening bodies, they must find individuals who have a will to join, or create such a desire where it does not already exist.

In the questionnaire filled out by members a number of

queries bore directly or indirectly upon the formation of groups. One of these read: "How did you become interested in the listening group to which you belong?" and another, "Which of the following purposes did you have in becoming a member of your group?" In the case of the latter question, the member was asked to indicate both his chief purpose, and any subordinate purposes or considerations which might also have influenced him.

In an Appendix three tabulations are presented which can be studied by any reader interested in full details. Here it will be sufficient to point out that the influence of clubs and organizations seemed to be the chief cause for arousing the interest of individuals in the idea of a listening group. Talk with friends or relatives, radio announcements, and notices in newspapers, in the above order of importance, were also "chief causes" for joining. The fact that a large number of group members came from P.T.A. bodies, or were recruited from Y.M.C.A.'s or community centers, undoubtedly helped to make the role of organizations an important one in the birth of listening bodies.

Naturally, the way in which an individual becomes interested in groups is only one factor in the growth of organized listening. His purpose in joining a group is an equally important consideration. This purpose was often associated with clubs or organizations. Of 336 members giving a chief consideration as moving them to join groups, 84 checked "to cooperate with your club or other organization." However, the desire for education or information was clearly a more powerful force—221 members checked one of the reasons falling under this general head as the outstanding force which influenced them to listen and discuss with others. An analysis of the subordinate purposes which were checked also shows educational considerations to have been the chief contributing factor as well as the main one. Educational interest in some form or other accounted for 631 checks in a total of 1,084. However, it should be noted that among the secondary purposes the following were important: "To enjoy meeting with people you liked" (169), "To save yourself by 'keeping up' through radio instead of reading, etc." (119), and "To get information more cheaply than you could in other ways" (48).

We might summarize the evidence on these two questions by

saying that organizations probably supplied an impulse to engage in group listening for at least a majority of all the members who reported, but that clearly the chief incentive was a desire to learn—in one form or another. The social aspect of group activity should also be noted as a secondary factor, and there is evidence that not a few members came into groups in the hope that information could be absorbed more easily in this fashion than by reading or attending classes.

In planning the members' questionnaire, I was influenced by the possible nonradio habits and opportunities of group members. For example, I wondered whether or not listening groups were made up of "joiners"—persons who would belong to almost any new and popular organization out of a confirmed habit of belonging. I asked myself, too, if some of them were persons to whom few or no other possibilities for self-improvement were open. Furthermore, I was curious as to what if any organizational activity members had given up in order to share in group listening.

Questions covering all these points were included. For example, members were asked if there were nonradio groups of an educational type to which they could belong if they desired. Of the 484 whose replies were tabulated, 299 replied that nonradio groups were available, 120 reported that none existed in their communities, and 55 did not answer. Members were also asked if they actually belonged to such nonradio groups. A total of 297 stated that they did, and some checked several organizations—382 checks in all were made. It was shown that the nonradio groups were chiefly evening classes (66 checks), discussion groups (65), young people's clubs (58), music groups (44), and drama groups (23). Ten members reported that they had left nonradio groups in order to participate in radio group listening.

These responses can be interpreted in various ways. More than 60 percent of the group members show that they felt they had other ways of getting information or instruction in addition to participating in group listening. All but two of those who reported having such opportunities had made some use of them. But almost exactly a quarter of all group members felt that they did not have other opportunties. It is also clear that even where opportunities existed, they were not always of the kind represented by radio groups; indeed, most of the nonradio organizations to which members belonged would not give much information about family life problems, and certainly art and music classes would not supply discussions on public questions! So to a considerable extent group listening seemed to be supplying a definite service not otherwise available—or at least not known to be available by the members who testified. Clearly, too, on the basis of the questionnaires, most group members were not "joiners"—if their own testimony is to be accepted. An average of much less than one nonradio organization per person indicates that on the whole their activities beyond the radio groups were limited.

Group leaders also gave testimony on nonradio activities. They were asked if their members could get comparable opportunities by joining nonradio groups or organizations. Thirty-five replied that their members could; 62 that they could not; 8 did not reply. This response checks pretty well with the report of individual members on their local opportunities.

In addition, the leaders were asked why, if opportunities in nonradio organizations were available, the members preferred radio group work. A list of possible reasons was offered for checking, with the following result:

Reason for Preferring Radio Group to a Nonradio One	Times Checked
Radio represents a more interesting way of presenting information	32
Radio offers a better device for starting discussion	28
Radio offers an opportunity to learn without doing much work	19
A radio group represents a novelty	13
A radio group is just another activity taken on because it is available	4
Total (44 leaders replying)	96

The testimony as to opportunities checks fairly well with my own observation. In many communities which I visited in the course of the Study there were clearly few nonradio discussion groups which members could have joined; in some communities, none. My experience with men's discussion clubs, gained in a study made in 1938 for the American Association for Adult Education, showed such nonradio bodies to be numerous if the totals for the country as a whole were considered, but to be distributed unevenly.[1] They were most plentiful in the larger cities, and I was impressed with the fact that their members were almost exclusively business leaders or professional men. The radio groups reached down into strata unable to afford or to have time for the fairly expensive lunches or dinners which, in the "regular" discussion clubs, were usually the preludes to discussion. Again, radio units did not demand that their members have an established position in the community, or in a business or profession, as the luncheon and dinner clubs usually did. Students, housewives, clerks, nurses, stenographers, even the unemployed were usually welcomed by the listening groups, the only qualities demanded being interest and courteous behavior. In fact, there were groups composed almost entirely of students and the unemployed. In other words, radio seemed to be giving new opportunities, and giving them in a more democratic manner than has often been the custom with other agencies in the past.

Calling All Registered Listeners

All that has been said in this chapter refers to listening groups and does not apply to groups of registered listeners. Yet they too must be stimulated to undertake their activity. How are they attracted to the programs which they hear?

Obviously they cannot be dealt with locally. They must in some way or other establish communication with a radio station or with those managing a program, and send in their names, together with the money for the fees which in some cases pay for the study aids which they receive. Quite naturally in these circumstances, the radio stations or those in charge of the programs in question have taken the initiative and have invited listeners to register. Invitations have been extended in various ways: by

[1] Frank Ernest Hill, *Man-made Culture* (American Association for Adult Education, 1938).

leaflet, by newspaper publicity, through public agencies like schools and libraries, and by radio announcement.

The latter method has apparently been the most effective. Of 423 registered listeners, representing altogether 16 programs, 406 made replies to the question: "How did you first become a registered listener?" Six possibilities were listed for them to check—"You were informed in another way" covering any possibility not specifically phrased by the other five. A listener could check several causes for engaging in his activity if he felt that more than one had existed. A number did so—altogether 490 causes were checked.

Of these 273 were set opposite the statement: "You heard by radio announcement that you could get study aids by sending in your name." However, 86 listeners reported that they had become interested through talking with a friend or relative, while 60 had been influenced by announcements received through the mails, 52 by notices in newspapers, and 3 by talking with their local librarians. There were 16 who had "been informed in another way."

Thus radio itself, in so far as this limited sampling is concerned, was the most effective means of interesting individual listeners in radio programs which offered opportunity for listening correlated with study. Naturally this fact does not mean that other ways of getting in contact with such listeners are unimportant, or even that some of them might not become even more effective than radio announcements. The latter are the cheapest and most convenient instruments in the hands of those broadcasting programs, and have perhaps been used more generally than circulars or press stories. Probably they are also more effective. At any rate, it is apparent that if the 406 registered listeners are representative of their kind, most of the latter have undertaken their special activity in connection with educational radio programs because of information they got from their loudspeakers.

With the registered listeners, there was of course no question of their being formed in small groups under the encouragement of an organization. Organizations could and did encourage them —Town Hall, for example, and a number of agents for univer-

sities. Still, the appeal had to be individual, and was made chiefly
to the listener's desire to improve himself. This situation is
reflected in the registered listeners' responses as to their purposes
in taking up radio listening. They checked the possibilities offered
them as follows:

Purpose of Listener	Number of Times Checked
To study	168
To improve a skill in some art, craft or hobby	91
Other purpose, not specifically listed	39
To save time by "keeping up" through radio instead of by reading, etc.	31
To see what education by radio was like	14
Total	343

Eighty registered listeners did not reply to this question.
However, the predominantly serious turn of mind which the
majority exhibited is clearly revealed in the above table.

A slightly larger percentage of registered listeners than of
members reported that there were no nonradio opportunities for
self-improvement of which they knew. About 27.2 percent answered to this effect. Moreover, their answers were sometimes
accompanied by revealing comment. "Radio was the only source";
"There probably are [nonradio opportunities] but I have not
been able to get away from home"; "There is no group here";
"Not what would interest me"; "I have youngsters and a home
to care for"; "Not near—live in country"; "The radio is my only
contact with the outside world for the present." Other registered
listeners indicated, as did many group members, that while nonradio group work or instruction was available, it was not what
they wanted, or was offered at a time which did not enable them
to profit by it. An examination of both sets of questionnaires
convinced me that the registered listeners had on the whole fewer
opportunities and less freedom of movement than had members
of listening groups. They seemed also to be in the main more
appreciative of what they were getting than the members; to an
extent, probably, because to get anything was more difficult for

them than for the members. Of course the majority of registered listeners were apparently able to avail themselves of nonradio activities, and did so.

It is clear enough that many influences play upon a variety of persons to bring a radio listening group into being. Beyond doubt the attitude and actions of public-spirited organizations are important factors in stimulating group activity. The character of radio programs is clearly important—groups gather only when a subject interests them. The social aspects of group activity also attract some individuals, and stimulate the forming of groups.

However, the two basic elements are perhaps the character of radio itself and the desire for self-improvement. Radio is new, it is convenient, it seems to be a timesaver, it can bring distinguished speakers to the smallest villages, it makes the starting of meetings easy, and furnishes matter for discussion. Organizations interested in listening groups know that the loudspeaker brings these advantages. So, more vaguely, do the men and women who are potential members of groups. And the latter realize, too, that radio will permit them to satisfy their curiosities, their desire for discussion, their thirst for knowledge, and will do so with a minimum of inconvenience and expense. From the marriage of radio and its advantages with the hunger for self-improvement the group comes into being. Without these factors it could not be born.

V · PLACES AND PERSONS

THE trees on the steep Kentucky ridges are ablaze, but not with fire that destroys. The forests stand smokeless in the Indian summer of the year. Gold, copper, and scarlet blend startlingly yet peacefully into a moveless autumn pageant. There is no sound save that of the engine of the automobile, nosing up breath-taking turns, which a moment later lie outlined below, like coils of rope carelessly dropped; or diving into slits of valleys where a few houses stand amid fragmentary cornfields; or even running along the strangely smooth bottoms of creeks—shale or gravel—for miles at a time. You motor perhaps one hundred and twenty miles from Lexington, chauffeured with acrobatic skill by Elmer G. Sulzer, director of the University of Kentucky's radio activity, before you come at last to a "center," with its school and living quarters and perhaps its church. There you join a group of children or adults, or both, gathered before the radio to listen to a chosen program and discuss it.

Such a group is among the most remote to be found in America. During many months of the year it can only be reached by horseback or on foot for the last several miles of the route, with plentiful wading through snow, mud, or water. Yet dozens of groups are functioning in these fastnesses throughout the cycle of the seasons, and to have a sense of where listening groups are to be found one must be aware of these isolated gatherings. For many groups are relatively remote. One becomes conscious of this fact if he takes a list of Iowa or Oklahoma or even New Jersey groups and circles with red on the map the hundreds of localities (if indeed they are to be found even in a good atlas or a Federal list of post offices) in which listeners gather to hear radio programs together. You can drive out a few miles from

McAlester, Oklahoma, and find a group at the state penitentiary. You can twist through the Massachusetts hills to join one in a CCC camp. You can go by train and bus and taxi to meet with another at a private residence in the outskirts of a New Jersey town.

From Church to Bar and Grill

On the other hand, it is possible to go, on a Sunday, by subway to the central Health Building in Manhattan and find a large group in this deserted downtown district. Or you can walk down the long main street of Highland Falls, New York, climb the stairs of the town hall, and meet with a group only a mile from the boundary of the West Point reservation. In residential Chicago, on dozens of college campuses, in the suburbs of San Francisco, in Brooklyn or the Bronx, similar gatherings are to be found. The geographical range and immediate physical setting are great and varied, respectively. Groups meet everywhere and in many kinds of quarters.

These places of meeting are important. Groups cannot exist in empty space, and if you want to picture them, you must have a sense of where they are and of the bulidings they use. There is no norm either for location or setting.

To be sure, a large number of groups meet in private residences of some kind. Of the 105 leaders who reported, 31 gave their meeting places as "private residence," presumably the same house being used in each case for all meetings. Others—33 more —changed their places of meeting from broadcast to broadcast, for they reported meeting in "various residences." Most of these 64 groups were P.T.A. units, some of them very small; probably not more than half of the groups in the country meet in homes. The other 41 reporting to the Study met in 20 different types of places—a total of 21 varieties for all the groups heard from.

Next to homes, school buildings were the most popular centers—16 groups used them. College halls of one kind or another served 5 groups, Y.M.C.A.'s another 5, community houses and churches 4 each. Then followed a miscellany of places reported by two groups or by one—Y.W.C.A.'s, Jewish community centers, radio stations, tea rooms, hotels, CCC libraries or radio rooms, town halls, village board rooms, clubs, health centers, and settlement houses. In Louisiana a group even met in a bar and service station! "The field representative of the University notified us of the agricultural programs," writes M. Blue Pinsant, proprietor, "and I notified my customers."

The mere enumeration of these varied meeting places gives little idea of their individuality. The residences, as discovered by actual visit, differ greatly one from the other. The little inn near Brightwaters, Long Island, a mile distant from the village on a motor highway and tended by sturdy German-Americans who invest it with an almost Tyrolean atmosphere, is different from a hotel in mid-San Francisco or Chicago. The memorial assembly hall of the Maxwell Graduate School of Citizenship at Syracuse University is an individual room unlike that used by college groups in Kent, Ohio, or Hopkinsville, Kentucky. One has to know group quarters personally in order fully to realize their varied character.

Three Kinds of Groups

The same is true of the numbers and appearance of group members. Statistics, to be sure, prepare one for variety as to the size of radio listening units. The more than one hundred groups reporting directly to the Study ranged in membership from 3 to 2,000 persons. There were a number of groups which gave memberships of 5, 7, 8, 11. There were also groups with 25, 50, 150, 200; and the second largest group reported 1,000 members. As previously stated, the average membership of the 105 on which the leaders reported was 58.4 persons, although if the 2,000 and the 1,000 member groups were omitted—and they seemed somewhat special in character—the membership average was about 29.[1]

---

[1] Since writing this paragraph, I have obtained precise data on the group reporting 2,000 members. It really comprises 24 units, and represents the Ohio County P.T.A. Council, with headquarters in Wheeling, West Virginia. The membership of the various units ranges from 50 to 400, according to the Council chairman, while the "average attendance"—presumably per group—is given as 50.

The group of 1,000 met in New York City at the place where a musical quiz program was broadcast, and gave its average attendance as 150. In the latter case the full membership represented all persons who attended "regularly" if infrequently, and, as might be expected where so many persons were involved, there was a constant change in the size and personnel of the audience from broadcast to broadcast, although the leader claimed a core of fairly regular attendants.

This number represents about the average attendance found in visits to groups. In no case were there fewer than 10 persons, and usually from 20 to 35; but the Syracuse University gathering numbered from 200 to 300, including students and adults, and this was clearly a usual number for that group.

In fact, both records and personal visits indicate that there are three types of groups as to size.

The small, intimate gathering is most characteristic of parent-teacher units. P.T.A. groups with as few as 5 members are encouraged in some states so that the discussion may be informal and intensive. There are nevertheless other kinds of groups—Town Meeting bodies among them—with 10 or fewer persons.

The medium-sized group of from 15 to 35 is probably the most usual. It parallels the informal, nonradio "dinner club" or "discussion group." It is rarely representative of the whole community—rather it is built up from a small nucleus of friends and their friends, although often an effort is made to bring in varied types of persons. Sometimes it represents a small suburb or village pretty fully, sometimes a neighborhood, sometimes a church or Y.M.C.A.

But groups formed in institutions like the latter often take on the character of a community meeting, as do those fostered by clubs. They thus represent a third type, much larger than the first two. The Town Meeting group at Allentown, Pennsylvania, is promoted by the local Women's Club, but reaches far beyond the Club for its membership, and usually is composed of from 80 to 120 persons. Groups meeting in town halls often make a conscious effort to draw in all the thoughtful people of the locality. They take on the character of forums. The meetings are considered failures if not more than fifty people attend; they are conducted with a formality and an expectation of drama which makes them occasions in the life of the community.

What the Leaders Say

As indicated by the above descriptions, size determines to some extent the character of a group. In general, the smaller the group the more homogeneous it is likely to be. Even when the group is large and the membership widely varied, the question still remains whether or not those who undertake group activity have certain qualities that set them apart from their fellow Americans.

Perhaps to an extent they have. In the main, they do not intend to be different; more than a majority of them would say that they were representative of their towns or villages. Many seek consciously for variety in membership. "I guess we have a pretty good cross-section of the community," is a remark that many leaders make. Even the small P.T.A. groups show a democratic desire to share their activity with all types of mothers; one has only to look at the faces of the women who assemble, whether in Oklahoma or New Jersey or Iowa (and, I am sure, in California or Texas or Michigan too), to perceive that there is wide variation as to income, education, and social background.

Yet not every citizen is serious-minded enough to listen to an educational radio broadcast, discuss its significance, and perhaps do a certain amount of reading before or after the broadcast. A visitor who has "listened in" with a variety of groups is likely to feel that all the people whom he has seen have a will to learn and that most of them have an intellectual curiosity or an interest in a particular subject. Perhaps this seriousness of interest is the most obvious characteristic of group members in general.

Yet to say this is to deal still in generalities, and we need not be so limited. We have rather extensive testimony as to the age, education, occupation, religion, political belief, approximate income, and even the state of mind of group members! The testimony is their own. Let us examine it.

As to age, 484 persons in 89 groups supplied fairly full information. The range for those who gave their ages—398 in all—is from 11 to 77. The average is 34. Members fall into the following age groups:

Age	Number in Group
Under 20	64
20 to 30	78
30 to 40	137
40 to 50	80
50 to 60	26
Over 60	13
Total	398

By age, the distribution is thus fairly even. The 30 to 40 group is the largest while about 69 percent of all members who reported are under 40 years of age. Youth and early middle age thus bulk large in the membership.

The education, occupation, income, politics, and religion of group members were reported by the leaders who were asked to make a guess as to these details. Sometimes they refused to do so. "Impossible to tell"; "We never discuss politics or religion"; "Private matter on which I have no information" were some of the replies. Altogether, 33 leaders out of 105 gave no data on political beliefs; 14, none on religion; 9, none on occupations 10, none on incomes, and only 4, none on education. Except in the case of political beliefs, the responses were thus pretty full. The question on income was phrased: "Please give your guess as to the number of members of your group who belong to families in the following income groups."

As to politics, the results showed that of the 5,905 members reported for, 1,104 were Republicans, 1,093 Democrats, 87 Socialists, 61 Progressives, 29 Communists, and 7 of "other political beliefs"; 3,524 were not reported on at all.

In religion, the groups were estimated by their leaders to contain 1,712 Protestants, 868 Jews, 615 Roman Catholics, 376 of no religion, and 7 of "other religions"—a total of 3,578, leaving 2,327 members not covered by reports.

As to occupation, there were 1,569 housewives, 1,176 students, 554 laborers, 503 professional workers, 400 office workers, 367 farmers, 212 salesworkers, 180 owners or executives, 139 unemployed or working on government projects demanding relief status, and 41 in other occupations. Only 764 were not covered by these returns.

Reports on income showed that the members were for the most part in moderate circumstances. Of the total, 1,404 were reported as probably from families with less than $1,000 a year; 1,540—the largest number—were put in the $1,000 to $2,000 brackets; 1,149 were supposed to be members of families receiving from $2,000 to $3,000 a year; while 520 were stated to be from those receiving more than $3,000 a year. These estimates covered a total of 4,613 members, leaving 1,292 not reported upon.

Educationally, the leaders made the following estimates: less than grammar school, 435; grammar-school education and part high school, 1,460 (the largest group); high-school graduates, 1,315; some work beyond high school, but no further degree, 972; college graduates, 1,242. Altogether these estimates accounted for 5,424 or 5,905 members, leaving but 481 for whom no report was made.

An effort was made to get a sense of the racial inheritances of groups. Leaders were asked first, if their groups were made up of persons with a single background (in addition to their American associations); and, second, if this were not the case, whether or not definite minorities—e.g., Czech, Italian, or Polish—existed within the group.

To the first question, eight leaders replied in the affirmative, reporting Negro, Jewish, and Polish groups. Twelve others stated that within the groups there were strong minorities, and specified Polish, Czech, Italian, Hungarian, German, Jewish, Negro, Swedish, Mexican, and French "Acadian" members. The majority of leaders either did not answer these two questions (17 and 38 respectively) or wrote "all American," "no such groups," etc.

These answers reveal that radio group listening was reaching a wide variety of Americans, more than seven percent of the groups having a special racial or national background, and more than eleven percent having national or racial minorities.

Autobiography of Members

So much for what the leaders say. The members themselves testified on a number of the same matters, and their autobiographical evidence deserves consideration.

I have already offered their reports as to age. They supply us also with data on sex. Here the record shows that women predominate. Of 477 members who gave their sex, 327 were women and 150 men. Nine did not answer. The presence of 50 groups out of 105 which met in the daytime and dealt with family problems might prepare us for this more than two to one proportion of feminine members. Until many more adult educational broadcasts during evening hours are available, this proportion seems likely to be maintained. However, groups meeting in the evening attract men. Questionnaires were received from 180 Town Meeting group members, and all but 3 answered the question on sex. The count shows 108 men, and only 69 women.

The differences between the leaders' reports and the members' as to occupation, education, and income are relatively slight. The four occupations most frequently checked were identical for both sets of questionnaires, and in the same numerical order. Housewives led; then came students, professional workers, and office workers.

The members reported about a year more of formal education for themselves than the leaders did for them; and they seemed to have a somewhat better economic condition. Some of the differences in the two sets of reports arise perhaps from a tendency on the part of leaders to distribute questionnaires among more energetic and intelligent members, who might be expected to average higher in schooling and income. Leaders were not asked to do this, but a number reported that they did. Also, members with better education may have found the filling out of the questionnaires easier than did the others, with the result that a greater proportion of them returned completed questionnaires. For such reasons, I am inclined to accept the leaders' estimates as probably more representative of group membership than the reports of the members themselves. Members were not asked to give their religious or political convictions.

The members gave further facts about themselves not given by the leaders. We have seen (from a consideration of why they joined groups) that about three-quarters of them reported membership in organizations other than their listening groups. We have seen that twenty-five percent reported that they belonged to no other organizations. Both these reports help us to get a sense of the kind of people who belong to radio groups. From personal contact with perhaps seven hundred of them, I can report that many have other group activities—activities probably more numerous than their questionnaires record. They seem normal citizens in this respect; most middle-class Americans belong to organizations, and many who would be termed below the middle class economically. I could understand also from seeing groups in action why many members reported no other group activity. Some clearly had none. They were young persons of little means, housewives in small towns, citizens in communities (yes, there are such in America!) not very well "organized." We may think of radio as bringing to educational activity, through its listening groups, a number of persons who have previously not participated as adults in such activity. Probably the percentage would be as high as twenty-five, perhaps higher. It should be remembered that about a quarter of all members reported having no part in other organizations.

Members were asked if they were newspaper readers; 425 replied that they were, and 56 that they were not; 3 gave no answer.

Queried also as to their marital status, 317 reported being married (including 4 widows, 1 widower and 1 divorcee), while 164 gave their status as single. Thirteen did not answer. It can be seen that the married predominate. Because of the presence of a number of student groups, the greater part of those reporting themselves as single are less than thirty years of age.

Members read, as well as listen to the radio. They were asked their preferences as to radio programs, magazines, and books. The replies brought in such a quantity of material that a full analysis was impossible, considering the time and funds available for the work of tabulation. Accordingly a sampling of one hundred questionnaires was made, distributed carefully as to the geographical location of members and the type of program heard. The following were the ten most popular choices reported by these questionnaires:

Radio Programs: *Information Please; America's Town Meeting of the Air; One Man's Family; Ford Sunday Evening Hour; Jello Program* (starring Jack Benny); *University of Chicago Round Table; Metropolitan Opera; Chase & Sanborn Hour* (Charlie McCarthy); *Major Bowes' Amateur Hour;* and *Lux Theatre of the Air.*[2]

Magazines: *Reader's Digest; Good Housekeeping Magazine; Ladies Home Journal; Life; Saturday Evening Post; Time;*

[2] Many general designations like "symphony," "quiz programs," "news broadcasts," and "good music," were made, but naturally these could not be tabulated. Had such designations been specific they might have changed several items on the list given above.

American Magazine; McCall's Magazine; Parents' Magazine; and *Cosmopolitan Magazine.*

Books: *Grapes of Wrath; Rebecca; The Yearling; Gone with the Wind; The Nazarene; All This and Heaven Too; Disputed Passage; Inside Asia; Escape;* and *Christ in Concrete.*

Who Are the Group Members?

Clearly the evidence presented above is fragmentary and to an extent subjective. Still, it gives some basis for a tentative characterization of the people in America who belong to listening groups. By the 105 leaders and 484 members whose questionnaires were tabulated, 24 different radio programs and 22 different states are represented. (A list of these was given in the Foreword.)[3] To be sure, almost three-fourths of the programs deal either with public questions or with family life problems. However, at least a majority of all American groups listen to these two types of broadcasts. Conceding that it would have been desirable to have more groups that heard musical or semieducational offerings, we still have something pretty close to a cross-section of group listeners.

What then can we say of these men and women coming together to hear and discuss radio programs? Certainly we can point out that they represent a broad range of natural and racial inheritance, and a variety of religious and political faiths, although neither Democrats nor Roman Catholics are represented so fully as their known ratio to the general population would lead us to expect. We can recognize, too, that the occupational range is wide among members. At the same time, we can perceive that there are relatively few farmers, laborers, or factory workers; "white collar" vocations predominate. Or rather, such vocations are in the family background of most members, for about forty percent of them are not employers or wage earners, but housewives or students.

In education, listening group members stand fairly high. The average American has had between eight and nine years of schooling; the average member of a listening group has had between twelve and thirteen. However, he is not a "high brow."

[3] Pages 12 and 17.

While in his group he may find a teacher or two, a few successful business men, doctors and lawyers, and often college students, he rubs shoulders also with nurses, stenographers, clerks, salesmen, mechanics, and persons on relief. His radio listening and reading, while often good in quality, are on the whole what might be called popular.

Conversations with group members indeed revealed the humbleness of the origin and the limitations of education of many members and even leaders. One of the latter, a woman in a Middle Western state, talked frankly about her early struggles to get an education. She had been an orphan. For a time she had worked as housekeeper for a family in the town where I found her, getting her high-school education at the same time. Then she had married. Her English was not of the best, but she was sincere and intelligent. As the mother of three children she became interested in a radio program dealing with family life problems, because it offered her counsel from trained psychologists and pediatricians. Two neighbors, women who were university graduates, urged her to form a listening group, and insisted that she lead it. She got the majority of its members by going from door to door and telling her own experience. No one talking with her as I did could have failed to be impressed by her integrity, and by the fine spirit with which better-educated women deferred to her and coöperated with her.

"I asked both Mrs. Stone and Mrs. Woodford [these are fictitious names] to be leader," she reported. "I told them I had little education and no experience in leading. But they said that I must lead. They would help—and they have—but I was the right one to start and carry on the group. So I did it. The other mothers have responded—well—just wonderfully."

This is not an isolated case. In another group I found the "spark plug" to be a young Italian-American clerk working in a warehouse. Teachers and doctors and bankers were encouraging him and taking orders from him!

If one must depict the "average" listening group member, one would say that he comes from a family with a modest income —it averages, one would conclude, less than $1,700. This is definitely above the American norm, but means that life must be lived with economy and often frugality. Naturally, average incomes vary from group to group. In many cases, the entire membership was reported as being from families with less than $1,000. In few groups were there many members from families with more than $3,000.

All in all, the members of radio listening groups are pretty much "plain folks." They usually have their civic and professional leaders, but they work in an atmosphere that is unpretentious, as the statistics indicate. There is nothing about most groups to make one think of them as socially or economically above the level of their communities—except the seriousness which they bring to their activity.

This, indeed, is distinctive. Everywhere one is impressed by the earnestness of many group members. This earnestness has been indicated already in the discussion of how groups come into being. I became aware of it in the course of my visits to various listening bodies. The groups engaged in discussion, for example, have usually drawn to them people who believe in the social importance of discussion. They feel that they are assisting to make democracy work. To be sure, they have been told so by Commissioner of Education John W. Studebaker, Mr. Denny of Town Hall, and others; nevertheless it is significant that they believe in what they are doing, and think carefully and keenly about it. In a similar fashion, one finds the P.T.A. groups aware of the importance of family relationships and eager to get authoritative testimony about it.

I encountered the latter groups first in Iowa, and I was impressed with the thoughtfulness and intelligence of their discussions. "These Iowa women are remarkable," I said rather naïvely to myself. "They organize and think and discuss with something close to distinction."

However, a week later I encountered quite as much sincerity and keenness and wisdom in Oklahoma, and began to realize that American mothers were naturally interested in their children and appeared to best advantage in discussing family problems. When I met with my first New Jersey group I expected to find it manifesting the same admirable characteristics, and was not disappointed.

Who Are Registered Listeners?

It might be supposed that groups of registered listeners are composed of persons much like the members of listening groups; in the main, they are. Of 406 out of 423 who reported their sex, 305 were women. By occupation, the housewives were the most numerous (223); professional workers were second (57); students third (17); owners fourth (16). These figures indicate that registered listeners are on the average a bit better off as to income than group members, although among them were also farmers, salesworkers, office workers, and 31 unemployed. The educational level of the registered listeners was higher than that of the group members; one-third were college graduates as against about one-quarter of those in the groups, and 153 reported that they had been graduated from high school and had taken some college or special work.

The chief difference, however, lay in age. The registered listeners averaged 41 years as against 34 for group members. Two hundred and fifteen of them were past 40. A number had clearly sought the radio as a convenient contact with the outside world, and some noted age or disability as a reason for doing so. However, any attempt to explain the marked difference in years between the two types of listeners would be speculation. Let us leave it that the difference exists. In the case of Town Meeting listeners, the contrast was marked. The registered listeners averaged 50; the members of groups 34—the average age for all group members.

If the group members were earnest, the registered listeners were apparently even more so. They took their educational radio intensely. Of the 423 filling out questionnaires, 224 reported that they listened by themselves, while 169 stated that they listened with others, and 30 did not reply. But when asked if they preferred to listen alone, 53 did not reply, and of the remainder 271 reported such a preference, while only 99 wanted to have company. Asked to give reasons for their views, many replied: "Can concentrate better when alone," or, "I can get much more out of the broadcasts when by myself." A total of 249 reported that they took notes during broadcasts.

The general radio listening and reading habits of the group members and registered listeners were much the same, with the registered listeners showing tastes that were a little less popular.

In Summary

What do all the above facts and impressions tell us? Certainly nothing that is mathematically exact, but something that is informing and useful. Let us set down specifically what seems to have been revealed with respect to group members (and, in the main, registered listeners) by data gathered through questionnaires and by impressions gained in the course of visits. The following statements seem to apply:

1. Listening groups have attracted a wide variety of Americans, representing various racial and national strains, and different political and religious faiths. Groups differ widely, however, in composition, some being homogeneous, others heterogeneous.

2. Listening groups as a whole include more women than men. However, groups meeting in the evening include more men than women.

3. Groups show a great range of membership as to income. Many members come from families with very small financial resources; a smaller number are "comfortable" economically. The majority are from families receiving from $1,000 to $3,000 a year. Group members are thus a bit above the average of Americans in this respect, but on the whole are in modest circumstances.

4. Similarly, group members have had more formal education than the average American, but the educational differences within groups are often great; and in most groups the member with little formal education does not feel out of place. Some groups show a relatively low average in years of schooling; a few a high average.

5. Group members are young, relatively speaking. Youth and early middle age predominate.

6. Group members are not "joiners," although three-quarters of them carry on other organizational activity. However, one-quarter do not, and many persons belonging to other organizations do not get from these what they get from a listening group.

The groups are bringing to perhaps half their members opportunities which they cannot get elsewhere, or would not trouble to get.

7. Group members are interested in their work, and have an impressive earnestness and a desire for self-improvement.

8. Except as to age, registered listeners are apparently much like group members, with a somewhat better education and economic position, and an even greater seriousness.

With this list of characteristics in mind, we may leave our consideration of the group member and his cousin the registered listener as persons. I trust that we now have a fairly clear idea of what kind of Americans they are. We may now turn to what they do. Their actions as described in ensuing chapters will of course cast further light upon their characteristics, for after we know a man's racial background, age, education, reading habits, and affiliations, what he does still remains the most revealing comment upon his character.

VI · GROUPS IN ACTION

I T IS 9:15 P.M. on a Thursday, and four persons are seated in the spacious living room of a residence in a New Jersey town about thirty miles from New York City. In fifteen minutes the broadcast of *America's Town Meeting of the Air* will begin. These four persons, and presumably at least a dozen others, will listen to it.

Nine-twenty. Several more people have arrived, but the one visitor present begins to fear that this will be a slim meeting. He listens for the doorbell. By 9:25 there are nine people present. "Don't worry," the leader reassures the visitor. "The front door is open. They'll pile in during the next five minutes. They always do." Sure enough, the group grows rapidly. More than twenty have assembled by the time the radio set has been tuned in and the crier is ringing his bell. Five more straggle in during the next several minutes. The group has begun its evening's activity.

This was my first experience with a listening group, and it was an experience that was to be repeated a number of times. Yet it was by no means general. On another occasion the group met in a large hall a full hour before the program was heard. A chairman called the meeting to order—there were several hundred present—and in turn introduced two speakers, one upholding the affirmative of the question to be discussed, the other the negative. Each spoke for twenty minutes, and when the actual program came the audience was already interested in the issue being debated, and eager to see what the nationally known speakers appearing in New York would have to say about it.

Ways of Beginning

These two meetings represented very different ways of launching a listening group upon its evening's work. They by no means represent the only methods in use.

Another group may meet half an hour before the time of the broadcast, and listen to a statement by its leader which gives enough background information to make the listeners feel not wholly unfamiliar with the subject of the program which they soon will hear.

Another group may take up the broadcast of the previous week, permitting questions and brief discussion, and then proceed to the program of the evening.

Still another may have a general discussion in which issues are sharpened and questions raised which will later have pertinence.

Another may have as a guest an outside "moderator" who will present a preliminary statement covering the topic with which the broadcast will deal.

To judge by the report of their leaders, most groups do not hold a meeting prior to the broadcast. Of 105 groups, 40 reported that they did, 63 reported that they did not, and 2 made no answer. From the 40, a variety of answers were given to the questions, "How long?" and "What do you do at this meeting?"

No more than twenty minutes is spent by some groups in preliminaries—9 of the 40 so reported. The others consumed from half an hour to "all afternoon" with activities before the broadcast.

Few groups were specific concerning these preliminary activities—that is, whether there were appointed speakers, or whether there was merely a general meeting presided over by the leader. However, various descriptions of the general character of the meetings were given.

Some of these follow: "Discuss public speaking," "short discussion of topics," "discuss last program," "discuss questions to be heard," "roll call," "discuss questions sent by forum director," "have a well informed speaker prepare group for broadcast by an historical and educational talk," "give a book review," "discuss a problem," "talk of past and coming programs," "reports re reading," "anticipate with aid of study guide the play—its setting, etc." (for a *Great Plays* group), "hear our local specialist," "discuss previous broadcast and the broadcast to be heard—sing

French songs" (this for a Louisiana group listening to broadcasts in French).

From these comments one can get an idea of the various uses to which these preliminary meetings are put. In a few cases the groups holding sessions before the broadcast had no discussion after it. Thus their only opportunity to discuss the program was at a meeting held later—a week, a fortnight, or a month after the program had been heard. Yet most groups which discussed the program previously heard also held meetings and discussion after as well as before the radio offering. Thus they dealt with each broadcast twice. Apparently they wished to give opportunity for the expression of any thoughts which had come to members in the interval between meetings.

I began the Study with the assumption that a preliminary meeting would seldom be held, and if held would be of limited value. I closed my experience with the knowledge that meetings before the broadcast were fairly common, and often contributed greatly to the understanding and enjoyment of the radio program. The success of a group is certainly not dependent upon a preliminary discussion; some of the best group meetings that I attended began with the radio program itself. Yet other meetings seemed to gather momentum and depth from beginning their programs before the broadcast, and the logic of a preparatory period seemed to stand out clearly.

Tuning In

We are now ready to consider the way in which the broadcast is handled. Do I hear a sigh of relief from some reader who is a bit bored with the everlasting variations which listening groups show? "Well," he mutters to himself, "there can't be any variations here. The broadcast is always the same!" Ah, my friend, it is too bad, but there is where you are wrong! So far as a group is concerned, the broadcast can vary a great deal.

Of course, in the first place, how well any broadcast will be heard—or if, indeed, it can be heard at all—depends on various matters such as static, station coverage (i.e., hook-up), and the condition of the radio set being used. In my visits to the Kentucky mountain centers it was about an even gamble as to whether modern science would triumph in the shape of good reception, or whether the sharp-backed ridges, a weak battery, or a faulty aerial would be victorious. A number of groups in various states which would like to function, are unable to depend on reception. At one P.T.A. meeting, a lone husband who had been brought in to keep me company—or who was merely there by accident— was tuning in for the ladies, and for several minutes produced a badly mutilated program. An efficient-looking woman finally wandered over to the instrument, gave one of its knobs the fraction of a turn, and nodded smilingly as the broadcast took on volume and perfect clarity. Thus it sometimes matters a great deal who presides at the radio set.

However, we may set such considerations aside, although 66 of the 419 members of groups who dealt with radio reception in their questionnaires (65 made no answer) reported having difficulties with it. The ability of group members to hear a broadcast was not what I had in mind when I said that the broadcast could vary. I was thinking of four different ways in which the same program can be presented to a group.

The possibilities are as follows:

A group may hear a broadcast together, picking it up by a receiving set from the nearest radio station.

They may hear the same program as a transcription which may be rebroadcast at a different time from a radio station or from some other source such as a public address system, or which may be played as a record.

They may hear the program read from a script by one person, or, if there are several speakers or characters in it, by several persons.

The members may listen separately to the program, and meet later to discuss it.

There is of course the further possibility of a complete rebroadcast (not from a transcription) at an hour different from that at which the program was scheduled, or even with different persons in the cast (if it is a dramatic program). However, this possibility has not been included in the four types of presentation listed above, because I have encountered no listening group which hears a program prepared in such a fashion. I do know groups using all those methods of hearing a broadcast which have been listed above.

To be sure, most groups get the broadcast direct from a station. Of 105 groups represented by the leaders' questionnaires, 93 heard their programs in this fashion. Yet one group sometimes heard the actual broadcast, and sometimes read the script. Seven always read the script. Two heard their program by transcription, one from a radio station and one from a public address system; and two groups listened to the broadcast as individuals, meeting later for discussion.

As a matter of fact, I could have had reports from more groups that read the program or listened individually, but preferred not to have too many of them represented.

It may be argued that groups which do not hear the actual broadcast are not really radio listening groups. I feel about such an argument much as I used to feel about people who asserted that what Walt Whitman called his poetry was not poetry. To me the important thing about Whitman was that his writings had much the same effect as poetry, and an effect which could not be ignored, whatever the writing might be called. And with groups, those that read broadcasts or listen individually serve the same function as other groups; moreover they are too numerous and active to be passed by in silence. They can of course be defended technically as radio listening groups, for that matter, since it is radio that gives them a chance to exist, since it is radio officials who encourage and guide them, and since all of them would hear the broadcast direct if they could.

That they cannot is also an important fact bearing on potential and actual listening group activity. It raises the question of what rights certain groups of listeners have, and of what can be done to serve groups that are unfortunately located.

I have said that all the groups I have encountered would prefer to hear the broadcast direct. I was somewhat surprised to learn that this was so, especially after I had attended meetings of groups hearing transcriptions and groups having the program read to them, which I thought were quite as profitable as the meetings of "regular" groups.

Five Groups for One Transcription

However, members and leaders assured me that they longed to be able to hear the actual programs as broadcast for the first time. Even at Syracuse University, where the broadcast, available by transcription, not only sounded exactly like the original but also enabled the group to meet at an earlier and therefore more convenient time, there was no sentiment for the indirect method.

"We know it's just canned stuff," one member declared. "Using a transcription is more convenient for us, but we wouldn't mind staying up an hour later if we could get the program right off the air."

"Oh, I think there's no substitute for hearing the real program," said a member of a New Jersey P.T.A. group which met at 2:30 P.M. and read a program actually broadcast at 11:45 A.M., a time at which the members felt they must be at home.

Nevertheless, the program heard by transcription has remarkable possibilities. If it loses a certain psychological "edge," it can be used and re-used in a remarkable fashion. Consider the case of the transcription of *America's Town Meeting of the Air,* used by the Syracuse group.

This is furnished by the local station, WSYR, which has made other arrangements for the 9:30 to 10:30 hour on Thursday evenings and does not carry the program then. At the time the broadcast is being received at the station studio, a local group sponsored by the Syracuse Peace Council meets there to hear it. The following Sunday, the station rebroadcasts the program from 1:00 to 2:00 P.M., using the transcription. The transcription then goes to Mexico Academy, a preparatory school about forty-five miles from Syracuse, where it is played on Tuesday night. On Wednesday it comes back to Syracuse, and is played from the radio workshop there to the Syracuse University group which meets at 8:00 o'clock Thursday evening. The transcription is then sent to Cornell University, where a group functions under the encouragement of the university librarian. From Cornell the transcription travels to Kent, Ohio, for use by the state college there. Negotiations have been under way for sending it thence to St. Lawrence University, but when I was in Syracuse these had not yet been completed, and the transcription came back to the Syracuse University radio workshop.

Thus this one transcription serves five groups, counting the one at the station which listens as the program comes in; there may now be six using it, with the possibility that a group or two meets on Sunday to take advantage of WSYR's broadcast on that day. The failure of the station to carry the program at the regular time has thus developed some by-products with advantages such as the use of the broadcasts at the regular time could not offer!

After the Broadcast

Let us now consider that we have sat with our groups through any preliminary meetings that they may hold, and through the broadcast or whatever substitute for the broadcast they may use. What happens now?

Clearly, most groups are going to discuss what they heard, if only in a brief fashion. Yet we cannot assume that all groups discuss the program at this time. Of the 105 who reported through their leaders, 7 stated that they held no meetings subsequent to the broadcast, and 7 did not report at all on this aspect of their work. It may be assumed that with any of these groups there was some talk among individuals, and possibly in a desultory fashion by the group as a whole; but there was no meeting. The program was either discussed at the next gathering of the group, or it was merely heard and not formally discussed at all.

In contrast, the remaining 91 groups engaged in a great variety of discussion. The ways in which they carried on such activity fell into four chief types, with some groups varying from meeting to meeting, and thus using two or more forms. The following tabulation therefore shows a total of more replies than the number of groups—91—concerned.

Anyone familiar with nonradio discussion clubs or groups will recognize that listening groups follow in general patterns which have been in use for years—in fact, long before radio became a factor in American life. And except for the use of the broadcast, listening group meetings are often much like other meetings in which a subject of general interest is explored.

Type of Meeting	Number of Groups Using It
General discussion under a leader	57
Brief remarks in turn by all members, with leader presiding	30
General discussion without a leader	24
Talks by several chosen speakers, followed by discussion under a leader	8
Other kinds of meetings	4
Total for all types specified	123

One might assume that with radio groups outside speakers or even speakers from among the group members would be less in evidence than with nonradio groups. To an extent this may be the fact. In some groups which I visited the only prepared speaker was the leader, and the radio offering was undoubtedly a substitute for prepared work by members, or for an outside guest.

Yet such practice does not seem to be usual. Rather the radio program seems to stimulate member activity and even the importation of outside speaking talent. A number of groups with which I sat had a leader, a chairman, and a special speaker, all from among the members. Others had guest speakers, sometimes in addition to the foregoing. Others, as I have pointed out, staged debates before or after the broadcasts, using either members or guests. It is interesting to note that 44 groups reported that members acted as leaders in turn, and 36 had had the same leader throughout the year, or throughout the existence of the group, while 25 made no reply to the question on the rotation of leaders.

The majority of groups depended on their own membership for speaker talent; 54 reported that they never had guest speakers; 35 reported that they had such visitors on occasion; 10 reported that they invited outside speakers regularly; 6 did not reply to the question: "Do you have guest speakers?"

How Often Do You Meet?

One important factor in the vitality and significance of groups is the number of meetings held during the year. Most leaders reported specifically upon this point in their questionnaires. It was apparent that the groups responding fell into three general classes—those listening all year around; those listening during the "season" (usually from October to May); and those listening for a few broadcasts only (seven or less). I have already discussed these types of groups in connection with P.T.A. units. The following table will show the frequency with which groups as a whole have met. The few making no specific report, with one exception of a unit meeting throughout the year, have been included among the totals for the groups having seven meetings or less.

Groups with seven meetings a year or fewer	19
Groups meeting for the "season"	
Meeting more often than weekly	5
Meeting weekly	49
Meeting fortnightly	7
Meeting monthly	9
Groups meeting the year around	
Meeting weekly	10
Meeting fortnightly	3
Meeting monthly	1
"When convenient"	1
Not specified	1
Total	105

It will be seen that 59 of the 105 groups reported meeting weekly, either all the year around or during the "season." Ten, in addition, met fortnightly. The five that met more frequently than once a week can be added to these to make a total of 72 groups of the 105 whose activity was intensive. However, a number of the groups that assembled monthly held excellent meetings, and made an important contribution to their community life. Indeed, when it is considered that the usual lecture or music series contains only from five to seven offerings, the radio listening groups seem remarkable for sheer amount of activity. This fact must certainly be borne in mind in considering their effect upon their members and upon the intellectual life of the nation.

All things considered, it may be said that listening groups

have shown a remarkable energy and a notable individuality and ingenuity in developing their meetings. Indeed, they might almost be called radio talking groups, for no one can know them extensively without feeling that they have greatly extended the discussion activity of Americans in most parts of the country. Many groups, probably most of them, would not exist except for the stimulus of radio. Most of the leaders might engage in similar nonradio activities; many members would not. Radio has lured them from inaction to listening, and from listening to personal participation in talk.

The particular type of meeting apparently has relatively little effect on the success of the group. Groups with differing procedures thrive and perform well. Probably the size of the gatherings and the character of the members have much to do with the customs followed. One type of meeting is possible in one community and not practicable in another. One type of membership does well with informal methods, another with more formal ones. A knowledge of varying group practices on the part of the groups themselves may promote certain features; time will tell that. Relatively few groups are as yet much interested in others; mutual interest will probably grow. As it does, the quality of group work should be improved. Yet that quality is in general already high. The groups for the most part are eager and industrious. Many have already invested their new activity with distinction. As radio programs improve (for many of them can and will) and leadership and member participation develop, the radio listening group may well stand forth as the most notable instrument for informing and stimulating Americans with respect to important areas of political, social, and aesthetic interest that the country has developed.

VII · BUILD-UP AND FOLLOW-UP

G ROUPS are born, groups meet. Those two facts, when explored as we have explored them, might seem to present
the cycle of group activity. Yet for the most part we have offered
a statistical record which assumes that, once born, groups continue to function by a kind of perpetual motion. We have not
asked how they gather strength for continuing a successful life,
either from within themselves or from without.

Such a question is vital to the matter of group existence.
Groups are often as delicate and difficult to foster as infants, and
a small book could be written about their care and feeding. We
shall have to be satisfied with a chapter.

There are groups which draw their vitality wholly from their
own membership and the program to which they listen. For the
present I shall set them aside and deal with those that reach
for outside help in the struggle for a prosperous career. When
we have noted the kind of help they seek and get, we shall understand the self-sustaining groups with little if any further comment; for these find within themselves the same essentials received by the assisted groups from outside agencies.

Why Groups Need Help

Groups that are assisted in important ways probably constitute seventy-five percent of the more vigorous listening bodies.
They ask help for one or more of the following reasons:

1. They lack experienced leadership.
2. They need instruction as to how to organize and plan for
   meetings.
3. Members feel the need of instruction as to how to participate in group discussion.

4. Their leaders, even if experienced, need information and guidance with respect to the particular subject of the broadcasts which the groups hear.

5. Many leaders and members want to see copies of what was broadcast.

6. Leaders and members want lists of books, pamphlets, or articles to read before or after the broadcast, with the idea of getting a fuller understanding of the subject.

7. Groups feel the need of advice as to how to publicize their activities, in order to get new members and to make their own membership proud of group accomplishment.

It will be seen at once that all these reasons for seeking aid lead to types of assistance which radio programs themselves seldom supply. We are thus considering the building up of groups for radio activity by supplementary services not in the field of broadcasting—either printed or mimeographed materials, for example, or personal guidance.

Obviously, as already stated, some groups with members experienced in discussion work can supply these supplementary services. However, to do so requires special effort, and thus even groups which could be entirely self-sustaining welcome the short-cut of regular assistance from a competent agency. As for the majority, it is doubtful if they could function successfully without help from the outside. Some have no members with any experience in group discussion or study. Others may have members with enough experience and ability so that a little help will be adequate. Still others might function well if they had self-confidence, but this they lack. For all such groups, the assurance of guidance and support from an agency which they trust means the difference between never attempting anything and the launching of an activity which surprises them by its exciting character and its success.

Who Helps to Build Groups?

In Chapter IV, I have listed the various agencies that assisted in launching groups. Any of these may also aid in building up strength for continued life, and a number of them do so.

The most important type of organization giving aid to groups already launched is the radio program headquarters. Forty-eight of 105 leaders reported getting assistance from such a source. The agencies mentioned the next largest number of times were government officials (cited by 9 groups), and club organizations (cited by 6). Only 74 of 105 leaders who returned questionnaires said that they received printed or mimeographed aids, 27 stated specifically that they received none, and 4 did not answer the question with respect to such materials. The fact that from 27 to 31 groups received no study materials should be borne in mind.

We can best go into the question of how groups are helped to thrive by looking at the work of several program headquarters diligent in ministering to listening bodies.

Town Hall Gives Aid

Among these, the Advisory Service of Town Hall, Inc., which issues study aids for *America's Town Meeting of the Air,* stands out for the character of its work. Its services to groups are on the whole the most comprehensive that have come to my attention. For the year 1939-40, twelve kinds of aid were listed by the Advisory Service as available to group listeners. They are described below in the phraseology of the annual report of Town Hall:

1. A copy of the new book, *Town Meeting Comes to Town,* by Dr. and Mrs. Harry A. Overstreet.
2. *A Handbook for Discussion Leaders,* telling how to organize a listening-discussion group and how to conduct this discussion.
3. One copy for each person in the group of *How to Discuss—Suggestions to Group Members.*
4. Town Meeting bulletin-board poster, 10½ x 12 inches.
5. Advisory service by correspondence dealing with the methods of obtaining local publicity, securing group members, utilizing local organizations and institutions, raising funds for dues, obtaining special speakers, etc.
6. One copy of each special advisory bulletin issued.
7. Twenty percent discount on all Town Hall publications.
8. Printed card, listing subject and speakers, for insertion in the bulletin-board poster (weekly).

9. An article describing the background and issues with a list of questions to stimulate discussion (weekly).
10. A selected list of readings on the topic (weekly).
11. A complete "Who's Who" of the speakers (weekly).
12. *Town Meeting*—bulletin of the program. One subscription includes twenty-six weekly issues, each of which contains the complete broadcast of "America's Town Meeting of the Air" of the previous week.

It can be seen at a glance that these twelve items represent a formidable array of aids. Town Hall charged a fee of $9.00 for the full set, $7.00 for any eleven items, or $5.00 if items 1 and 12 were omitted. No other program headquarters offered so extensive a service, and none made a comparable charge. In addition, while it did not promise to send its officials to groups as a part of the assistance rendered, the Advisory Service did have its representatives visit a certain number of listening bodies, who were thus able to get personal attention in addition to whatever help was given by correspondence or through standardized aids.

Three Other Services

Before discussing this one comprehensive service in relation to the needs of groups, let us look at three others which, like the Town Hall's, have proved effective in practice. All these are offered to groups which listen to programs dealing with family life problems. One comes from the program headquarters of the New Jersey program, *Homemakers' Forum;* another is directed from the Family Life Institute at Norman, Oklahoma, and serves groups which hear the *Family Life Radio Forum* broadcasts; and the third has been developed by the program headquarters of the Iowa program, *Radio Child Study Club.*

All these services to groups represent considerably less in quantity than do the aids offered by Town Hall, but they parallel some of the chief items of the latter. For example, all send advance lists of programs, all provide bibliographies for outside reading, and all furnish leaders of groups with suggestions for the conduct of meetings. In addition, all stand ready to correspond with groups about their problems, and to give advice as to how these can be met.

Representatives of the three program headquarters just noted also visit groups personally. The Oklahoma and New Jersey programs are particularly active in this respect. Since they cover a smaller territory than does the *Town Meeting of the Air* (although all three reach beyond the confines of their states, and *Homemakers' Forum* has a network coverage) they are able to do more than Town Hall officials in the way of talking with leaders and helping to organize groups. All have a somewhat closer tie with state organizations like the P.T.A. than has the Town Hall with any single association supporting its program, and they can get the experience, prestige, and official assistance of congresses of parents and teachers and of other similar organizations. None of the three programs is designed exclusively for P.T.A. groups, and all of them get considerable support from other than P.T.A. sources.

Dr. Alice Sowers in Oklahoma, for example, has promoted the *Family Life Radio Forum* as a program for all the families of her state—indeed, "for all individuals and groups interested in education for family living." Five organizations have coöperated with her: the Education Division of the Works Progress Administration, the Extension Division of the Oklahoma Agricultural and Mechanical College, the Home Economics Division of the State Department of Education, the Oklahoma Congress of Parents and Teachers, and the Oklahoma Division of the American Association of University Women. From all these agencies have come groups for the program, while others are to be found in CCC camps, NYA centers, Y.M.C.A.'s, and even in the state penitentiary!

In New Jersey, *Homemakers' Forum* is administered from the state college by the Agriculture and Home Economics Division of the state Extension Service, and county demonstration agents are therefore the direct representatives of program headquarters. While they may find other forms of activity better suited to certain groups of women, they are promoters of radio group listening, and help to organize and assist groups. I met several of these agents and went with them to group meetings. To judge from what I saw and heard, they represent a valuable action arm for those who plan for and promote the program.

Give Us Leaders

Let us now go back to the groups. We have seen that they turn to program headquarters or other agencies in order to get certain help which is necessary to group vitality. The first type of aid listed was trained leadership. What assistance do groups get with respect to this?

They do not get enough, on the whole; because leadership is undoubtedly the greatest factor in group vitality and the hardest to create where it does not already exist. Seldom does it exist in fully satisfactory form.

Unquestionably the Town Hall Advisory Service gives the fullest help to leaders. Its handbook goes ably and pretty fully into the problems and possibilities of leadership. If leaders can be made by pamphlet, this little guide is likely to make them. In comparison, instructions from other program headquarters are brief and superficial. Town Meeting group leaders can also get help from items 3, 9, 10, and 11 on the Town Hall list. Item 9, with its lists of questions designed to stimulate discussion, is almost universally used by leaders. Other program headquarters supply such service, as already indicated.

Such assistance is obviously the making of many a group. Particularly does it fortify leaders who have some experience but who lack self-confidence. Definite information is given on what to do and how to do it, and specific material is offered for each meeting. Often this is sufficient to start a group which gathers experience and ability as it goes along. The radio broadcasts lift some responsibility from the leader's shoulders. He is not thrown entirely on his own resources, but has something to work with. As a result, his job is easier than that of someone leading a discussion for which information and ideas and issues must be developed entirely by the group itself.

Nevertheless, good leadership cannot always be successfully fostered by printed or mimeographed aids, fortified by personal correspondence. It often fails to develop even where rather frequent visits by headquarters officials are possible. In 1935 those in charge of *Homemakers' Forum* reported that "the problem of securing competent leaders is one of the greatest." There had been complaints as to changes made in the time of the broadcast, and such changes were thought to have lost the program many groups. "But," said the director, "it is not fair to leave the impression that changes in time have been the only causes for a falling off of the groups enrolled, for poor leadership has certainly taken its toll also." In Oklahoma, Dr. Sowers has sought funds for the training of leaders, and New Jersey has actually held institutes for such a purpose. The director of the *New England Town Meeting on the Air* also discussed with the writer the possibility of leader-training conferences. There seems indeed to be no question that adequate leaders are lacking, and that a number of them could be developed by short courses if such an activity could be financed. It seems certain that the extension of leader training would mean greater vitality to many radio listening groups, and would even prevent a number from perishing. I shall have more to say about this in ensuing chapters.

For the present, the point to be made is that assistance rendered from program headquarters and other sources, chiefly in the form of correspondence and the sending of aids, does encourage leaders to function and assists them greatly to discharge their duties. Leaders who answered the question concerning the value of such aids clearly felt printed materials were of valuable help. Of 105 leaders, 55 reported that these materials were "of great use," and 20 that they were "moderately useful." Thirty made no reply to the question. It is significant that although a third possibility, "Of little or no use," was provided for checking, no leader checked it.

In addition, 53 leaders took the trouble to comment on why materials had been helpful to them, and it was clear from these replies that the help of outside agencies was counted upon and often highly prized as a resource for stimulating talk or even as the basis for the entire discussion.

Help for the Members

The seven kinds of assistance wanted by groups with one exception touch the matter of leadership. Or shall we say that all seven do? For the feeling of the members that they need to understand their function in a group—the sole exception I had in mind—is indirectly related to leadership. As his members become used to participating, the leader finds his work much easier.

We have already discussed to some extent other needs than the need for experienced leaders. We have touched upon the desirability of having information on the specific subject of the broadcast and upon the planning for meetings. Inexperienced persons thinking of starting a group find guidance in such matters highly important. Leaders and members alike probably make themselves familiar with whatever instructions are provided as to conducting meetings, publicity, dues, the questions sent out for each broadcast, and such matters.

In fact, as has been previously pointed out, in the case of 44 groups there were several leaders, and leadership passed from one member to another.

It is also interesting to note that when leaders were asked with respect to printed aids sent to the group, "Are these for the leaders only?", only 27 replied that they were, while 47 answered in the negative. To the next question, "If not, are they for all the members?" 43 leaders answered in the affirmative, and only 6 in the negative. (Thirty-one leaders failed to answer the first question, and 56 did not reply to the second. It will be remembered that from 27 to 31 groups did not get aids at all.) Thus a number of the group members seem to have shared in the building up of the majority of the groups.

The aid used most frequently by the groups with which I actually met was the list of questions sent out on the subject of each broadcast. To have these questions seemed highly important to them. They felt that whatever the particular subject of a broadcast might be—and sometimes it was regarded as one on which they needed enlightenment by experts—they had in these questions a definite basis for discussion. These showed exactly what should be dealt with. The replies of leaders to their questionnaires indicated that they also regarded these questions as plotting the course of the meeting for them—although I did not get the impression that the questions were followed slavishly. Certainly there was no disposition to deal with them too literally at any of the meetings which I attended.

As for reading lists, these were undoubtedly used, especially by leaders or assistant leaders—for sometimes several persons assisted with the discussions. At one meeting which I attended several members reported on reading which they had been asked to do.

Undoubtedly most groups have derived considerable benefit from correspondence, and some from the visits of officials representing the program. Everything considered, the printed or mimeographed aids and the contacts with those who supply them have done much to develop groups. Probably their chief value rests in the fact that they give the listening bodies simple directions as to how to take up an otherwise alarming activity. The directions say: "This is how you do it!" and even when explanations are made only in the briefest fashion, the groups are usually given courage to carry on. Indeed, the process of holding a discussion does not sound difficult, and with enthusiasm and a little previous experience it does not prove to be so.

Help from Home

At this point something should be said about the assistance provided by local organizations interested in groups. Undoubtedly such assistance is highly important.

It is varied in character. It may be represented, as previously stated, by the furnishing of quarters. A church, a college, a CCC camp may provide the group with a room. However, with the quarters there usually goes some advice from the minister, Y.M.C.A. official, or camp adviser. Perhaps such an individual acts as leader, or does so until leaders have been developed from within the group. Librarians also give assistance to groups in finding material, advising leaders, or advertising the activities of the group. To judge from the reports of leaders, libraries have not been as active as I had expected they would be. Only 35 of 105 group representatives stated that they or their members had been assisted by librarians. Forty-nine reported that they had not been assisted, while 21 made no reply to the question. Twenty libraries were reported as posting lists of books, 15 as advising leaders, 13 as advising members, 10 as rendering various other services, and 5 as providing meeting places. Leaders reported help from teachers (21), from newspapers (46), from town officials (6), and representatives of business organizations (7).

Sometimes the local sponsoring agency renders much fuller assistance than I have indicated above. The Syracuse University group, for example, always provides a faculty chairman, and faculty or other speakers for a preliminary discussion. The Allentown Women's Club takes full responsibility for filling the chair at its meetings and for procuring an outside moderator for every meeting. I can testify personally as to this, for I was drafted as moderator! The Jewish Community Center at Washington, D. C., always supplies a leader from its staff. In the Kentucky mountain centers the teachers or directors of the center almost invariably preside, and furnish an excellent type of leadership.

I remember arriving one noon at a little center in Kentucky. Our visit was unexpected, but the director volunteered to organize a group demonstration for us. The only available broadcast was one on Kentucky agriculture. I had heard other programs in this series and knew that the fifteen-minute offering would probably be dull. There were perhaps twenty-five adults and children, mostly the latter, in the room where the radio was placed, the children ranging in age from ten to fifteen years. I settled myself for a rather dreary experience. The program ran true to form. It was about sheep, and bore a heavy load of platitudes and statistics. I wondered what could be done with it.

Then the director stepped forward. "How many of you have sheep on your places?" he demanded. Hands shot up in various parts of the room. "What's a sheep for? What do you get from it?" he asked. "Wool!" "Meat!" "Coats!" came the responses. And the discussion was off to as lively a course as one would hope to find. Soon the boys and girls were arguing as to how their land could be adapted to sheep, why sheep had particular advantages for mountain farms, and who would supply information about them. I sat amazed. In a few minutes I was fully convinced that almost any radio broadcast could be used as the basis of a good discussion if there was a resourceful person to act as leader.

I have said nothing as yet about the influence of the P.T.A. in building groups. It is probably the greatest which exists. In the first place, every local branch, unless newly launched, has had experience in holding meetings. Leader-material may not be of the best, but it is available. In addition, the heads of councils and even the state officers have an interest in groups, and provide advice and assistance which is sometimes highly valuable. Programs dealing with family life problems lean heavily on P.T.A. members and officers, and usually get excellent support from such persons.

Servicing the Registered Listeners

The building up of groups of regular listeners is simpler than the developing of regular listening groups. As we have seen, such listeners are usually attracted by a radio announcement, by actually hearing the program, and by the invitation to write in for study materials. They become regular supporters of a program for the simple reason that they like the broadcasts and the materials sent to them so that they can make more intensive use of what they hear.

Some of their remarks about their experiences as radio students are so enthusiastic that they deserve quotation.

"I turned the dial and there was Professor Mott," writes an Iowa woman of less than grammar-school education, with regard to several of WSUI's programs. "Just like sitting down and talking with him. It was the same with Professor White. I had read, or tried to read, the story of the siege of Troy, and found it bewildering. When I tuned in on Professor White, I heard and almost saw ships dragged up on a sandy beach while the invaders 'dished it out and took it,' in the course of battle; all this he told in a voice, more like he was trying to soften a horrible scene, yet as he talked I imagined horses stepping in pools of blood as they drew along a blood-spattered chariot."

"Has increased both physical and mental ability!" says a Colorado executive about the calisthenics program, *Early Risers Club*, broadcast from KLZ in Denver.

"To my way of thinking exercising by the radio alone is better because I myself do better if no one is watching," a housewife says of the same program.

A listener to WOSU's French program claimed to have made

more progress by studying in connection with the broadcasts than her daughter had made in a French class in high school.

"Experienced people have been my 'teachers' over this radio system," writes a listener to *Columbia's Camera Club*. "They know what they are talking about."

A nurse in Los Angeles, hearing the same series and comparing it with a classroom lecture, writes "One is more attentive to a radio program because you can't ask the 'prof' to repeat."

Follow-up

I have referred thus far to forces which develop the group. A further aspect of group activity comprises the work undertaken by members as a result of group listening. It is the following up by members or leaders, or by both, of whatever stimulus and information they get from the hearing of a program.

The most obvious result of group activity is the reading which it may stimulate individuals to do. Leaders were asked a number of questions with respect to this activity. In reporting as to whether or not assigned reading was undertaken by group members, 39 leaders stated that it was, and 56 that it was not. Ten failed to reply. Eleven reported that all the members of their groups read assigned material, while 29 reported that some members did so. Twenty-eight stated that the reading was done in advance of the broadcast, while 20 reported that it was done afterward, "as a follow-up." Thirty-two leaders said that members made reports on reading to the group.

These figures, of course, refer to assigned reading, and not to that which individuals may have done voluntarily. The group members were asked a number of questions touching upon both types. Asked if they had done reading in connection with their group work, 293 replied that they had, 143 that they had not, while 48 did not answer. Asked to specify the type of reading most frequently done, 76 reported using magazines, 46 books, 29 pamphlets, and 23 copies of broadcasts. Other kinds of materials checked as used less frequently were magazines (149), books (131), pamphlets (130), copies of broadcasts (89), outlines (30), and other types (74).

Various reasons were given as to why this reading was done.

An interest in the subject led with 202 checks, next came the fact that reading had been assigned to members as individuals (74) and next that material had been mentioned in a broadcast (62). Thirty-nine members reported doing reading because it had been assigned to the entire group—making a total of 113 cases of reading done by assignment.

Of the 293 members who reported affirmatively, 277 answered a further question as to the source of their reading materials. As might be expected, some used several sources; so that altogether 477 sources of reading materials were checked by the 277 from a list of seven sources, including a category "other sources— please state." The local library proved to be the most popular bureau of supply: 148 members checked it. A total of 117 checked "books or other materials which you bought, or borrowed from friends"; 104 checked "other sources"; 47, "radio program headquarters"; 23, "radio station"; 5, "library package service"; and 3, "library wagon service."

Answers were made also to the question, "If reading has been done, has it been (please check) for every meeting, only for some meetings, or rarely?" A minority of 39 members checked the first of these possibilities. The great majority, 190, checked "only for some meetings," and 50 indicated that they had read "rarely."

Naturally, in making the questionnaires, the possibility of changes in reading habits was taken into consideration. Clearly these might be both positive and negative ones. Reading might increase in quantity or improve in quality through listening group activity. On the other hand, radio listening might conceivably be substituted for reading, or a change might take place such as the abandonment of better novels and biography for magazine articles and digests of information—an alteration which some educators might feel was of doubtful value, or even a definite deterioration. An effort was made to discover exactly what changes, if any, had taken place.

For example, several questions were included with respect to newspaper reading. The members were asked if they were newspaper readers now, and if they had been prior to joining the group. These questions indicated no change in habits, as 425 replied that they now read a paper regularly, and 426 that they had done so formerly! However, members were further asked if they read their newspapers more carefully since joining the group, less carefully, or with about the same care. This question revealed a belief on the part of 132 members that there had been a change to more careful reading, while 274 reported no change, and 20 failed to answer. Of Town Meeting group members, 56 out of 162, or more than a third, believed that their newspaper reading had become more intensive.

Another question, or group of questions, referred to the reading of books and magazines. Of the full 484 members, 401 replied to the question dealing with books, and 377 to that about magazines. A total of 117 felt that there had been a change in the first case, 131 that there had been one in the second. As to the kind of changes that had taken place, few members were specific. Altogether 14 reported an increase in the amount of book reading done, and 23 an increase in the amount of their magazine reading. Two in each case reported a decrease. Twenty-one members, however, felt that since joining their groups they had read more serious books, and 24 that the character of their magazine reading had become more serious. A number of members volunteered comment to the effect that they now read with more interest than formerly. There was a general evidence on the part of those who testified to a change that they had been stimulated by the broadcasts they heard.

Finally, as to the question, "In order to do reading for the radio listening group, have you given up other kinds of reading which you had been doing?" the response was "Yes" for 30 members, and "No" for 362, with 92 failing to reply. The few who had given up other kinds of reading reported for the most part the abandonment of light fiction, or the "channeling" of reading to bear upon the particular subjects of the broadcasts that were being heard.

Radio Is a Stimulant

In talks with leaders and members, I sought to get an impression with respect to reading which might act as a practical check on the returns from the questionnaires. What I found tallied fairly well with these returns. It was clear that in many cases the leaders "boned up," as did members assigned to lead the formal discussions or debates before or after the broadcasts. Apparently some members—often at least half of the group—scarcely read at all. This proportion seemed to vary. A number of members seemed to do reading sporadically, as their interest was particularly aroused. A number told me that they would see a book or magazine article dealing with the same subject as a broadcast they had heard, and would be moved to read it. Others spoke of their keener general interest in topics with which the radio programs dealt. One member of a Town Meeting group in New Jersey spoke positively of his newspaper reading.

"I read the foreign and national news in my newspaper regularly now," he said. "I've become interested in it. I used hardly to glance at such news—just read feature articles, personal stuff, and sports. Belonging to the group has certainly made the newspaper a different thing for me."

I should say that such cases were relatively few, yet the quickening effect of the group on reading seemed to be felt in some degree by a respectable proportion of group members, especially those listening to programs dealing with public questions. Doubtless an impulse was felt by more than a majority of most groups. The members themselves said it was. To a question as to whether the group made them want to "read, work, or study" or made them feel that they had got enough from the broadcasts and discussions so that they did not need to study further, 405 out of 434 who replied said that they had had an impulse to follow up their group work with reading or some other activity. Thus only 29 might be assumed to be using the listening group as a substitute for reading. Some specifically stated that they did. "I get just what I want from radio," said one. "I have very little time, and can't cover much ground by reading. What I get from the group saves me the effort of trying."

Everything considered, the group listeners seem to be definitely stimulated to read, and a few to read considerably. The majority read a little—sometimes only the mimeographed broadcasts already heard, sometimes only digests of information, sometimes magazines and books. Another minority do not read at all—or are not conscious of reading—as a result of group work.

It will be remembered that 143 of 484 group members put themselves in this class, and from personal observation and talks with individuals I should say that this was a fairly honest response. Such members probably read little anyway; a surprising number of Americans depend on their ears rather than their eyes for information. Probably a few more members than indicated by the questionnaires really use radio listening as a substitute for reading. Some of them do not wish to acknowledge the fact, even to themselves. Yet even assuming this to be true, the influence of group listening seems to be a positive one as to reading, although I feel that the strength of that influence ought not to be exaggerated. It seems to be only moderate in character.

A Springboard for Action

Yet reading is not the only type of "follow-up" which may result from radio group listening. For example, a broadcast and a discussion may make an individual think seriously about the subject covered, with the result that he forms an opinion or sharpens one already formed. Members were asked if the effect of group work was to make them reflect on the matters they had heard discussed, and 407 answered in the affirmative. A resolute minority of 39 reported no thinking whatever, and 38 did not answer the question. While such an inquiry undoubtedly held the door wide open for the entrance of wishful thinking, one who has visited a number of groups will I think be inclined to feel that the above quoted response had a strong foundation of truth in it. The groups do quicken interest and stimulate impulses for self-improvement, many of which never blossom into action.

If thinking is a dubious measure of effect, the doing of actual deeds is certainly less so. Even talking about the programs a member may have heard is more objective than mere thought. A total of 404 persons reported that they talked with nonmembers about the broadcasts they had heard, while 353 said they urged others to tune in on the series. Again there were minorities who neither talked (44) nor urged (26).

A list of possible acts which might result from group listening was offered to the members, and they were asked to check any of these which they had performed. Altogether, 283 members indicated that they had done one or more of the things suggested. The results were as follows: attended lectures, 164; used ideas or information got from group meetings in daily living, 151; saw exhibits, 35; carried on activities (such as gardening, playing a musical instrument, etc.) 30; and visited museums, 29.

When They Listen Alone

The data obtained from members of groups is in the main paralleled by that from registered listeners.

In the questionnaire framed for the latter, the questions on reading were slightly simplified. Again, the question about thinking was listed with other possible acts the listener might have performed as a result of his contact with radio. On the other hand, in the case of members, the reading of broadcasts, outlines, and other aids supplied by program headquarters or other comparable agencies was included under the general head of reading done; in the case of registered listeners, study aids were treated separately and the question on general reading was phrased, *"In addition to using study aids . . . have you done reading on the subject covered by your radio course?"*

Let us first take up the use of study aids. A few programs for registered listeners sent none of these (for example, *Columbia's Camera Club* and *Greek Epic in English*), so that such materials were not available to all the 423 listeners whose questionnaires were tabulated. However, 361 reported receiving various types of material to be used in connection with listening. Outlines led with 270 checks, next came pamphlets (170); then followed in order, lists of books (132), notes on art (58), pictures (52), other materials (45), and notes on music (30). The total of aids received was 659 for the 361 persons replying to the question.

Naturally, these aids represent a considerable amount of reading, and presumably they were used by all the individuals receiving them. At least, they reached those individuals, in contrast to the aids sent to groups, which often got no farther than the leader. However, the registered listeners reported doing other reading also—285 out of 423 testifying that they did, while 80 stated that they did not, and 58 made no reply to the question.

The chief sources of materials for this reading were "other sources," 165, and "local library," 158. An examination of comments volunteered in connection with the checks made for "other sources" shows that a large number of listeners bought or borrowed such reading materials. "Bought books," "Read my own books and magazines," and other similar remarks appear frequently on the questionnaires.

As with the members of groups, the chief cause for reading given by registered listeners was a general interest in the subject. Altogether, 237 persons checked this item. "Mentioned in the broadcasts" came next, with 146 checks, then "On a reading list," with 88. The total number of listeners who checked any source of reading materials was 275, and altogether 497 checks were made.

In order to cover any effect radio might have upon the kind of reading done, the registered listeners were asked in a single question if there had been a change in their perusal of books, magazines, or newspapers, and to describe the change or changes that might have taken place. A total of 165 persons reported changes. Of these, 56 stated that their reading of books had increased, 37 that they did more magazine reading, and 27 that they read newspapers more than before they became registered listeners. Seven persons reported decreases in reading: books, 2; magazines, 3; newspapers, 2. In addition, 89 reported changes other than those in amount—mostly changes in kinds of reading done, or in attitude toward reading.

When the responses of group members and registered listeners are compared, the differences are not great, but indicate that the registered listeners were definitely more active. A slightly larger proportion of them reported reading in addition to the aids they received than the proportion of members reporting on all types of reading, aids included. (All the 361 registered listeners mentioned above presumably used study aids, while 285 of 423 did additional reading. Only 293 of 484 members did reading of all kinds, including some study materials.) Almost the same proportions of members and registered listeners specified the sources of their reading materials—277 of 484 as against 273 of 423, the registered listeners again being more numerous in relation to the number filling out questionnaires. However, the members of groups made more checks for sources—477 as against 332. Even when radio stations and program headquarters, which sent only study aids, are deducted from the figure of 477, a total of 407 checks remains. The larger number of checks made by the members, however, may be accounted for by the fact that the group members were more likely to have a variety of facilities at their command than the registered listeners, a number of whom resided in the country, or in villages.

As to activities in addition to reading, 371 of 423 registered listeners stated that they talked about the programs with other people, and 326 asserted that they urged such persons to listen to the broadcasts. Here again the solitary listeners were somewhat more active than the group members. As to specific acts, 260 registered listeners reported that they thought about the subject of the broadcast; 142 reported that they utilized skills acquired through listening or applied what they had learned in daily living; 137 said that they carried on activities related to the broadcasts (speaking French, photographing, etc.); 89 that they attended lectures; 59 that they visited museums or art galleries; and 41 that they saw exhibits having some relation to the radio programs they had heard. A much larger proportion of registered listeners set down such activities than was the case with members —350 out of 423 as against 283 out of 484. However, it must be remembered that thinking about the subject of the program was included in the list offered registered listeners for checking, and was not on the comparable list checked by members. If we deduct this item, we find a total of 468 checks made by the solitary listeners, while the total of checks made by the members was 409. Thus the activities of the registered listeners, as reported, were not only more numerous, but also still greater proportionally, as the number of persons completing questionnaires was smaller.

For both groups, the reading and other activities reported as flowing directly or indirectly from radio activity indicate a definite stimulus coming from radio. For a few, whether group members or individual listeners, radio apparently became a substitute for reading. With the majority, it promoted not only reading, but thinking, outside discussion, and a notable amount of action. Another minority seemed scarcely to be influenced at all as far as reading or doing were concerned. We should not exaggerate the results of group work or solitary listening, but we can recognize that on the whole it seems to promote both mental activity and a tendency to do various things calculated to increase information, skill, and even wisdom.

VIII ▾ DISEASES AND MORTALITY

I F GROUPS must struggle to be born, if they must often work
hard to achieve vitality, it would be strange indeed if many
did not languish and die. Many do. Often groups are as difficult
to raise as turkeys in eastern Oregon or apricots in New Jersey.
Indeed, early in the course of the study I was told that most
groups were fly-by-nights, that impermanence was of their essence.
From the beginning I looked for signs of disease and death.

These proved to be less numerous than had been prophesied.
Nevertheless, abundant evidence came in with respect to groups
that were languishing or had failed. Let us look briefly at the
causes for their morbidity or death, and see, if we can, how these
affect listening group activity as a whole.

Groups fail for a number of reasons, all of which may be
thought of for convenience as falling into two main categories—
those that are external to group life, and those that lie in the
groups themselves. Sometimes, as one might assume, the two
sources of mortality are blended.

Acts of God

Many groups perish for reasons over which they have no control—which for all they can do are as inevitable as those catas-
trophes which the law names "acts of God." These events may
wipe out dozens or even thousands of groups at a time. They are
obvious and sometimes even desirable.

The simplest and most devastating of these is the termination
of a radio program to which a group or groups have listened.
Theoretically this need not mean the death of a group, for a
rugged listening body might preserve itself by shifting to a different radio offering in the same field, or even by turning to one

beyond that field entirely. For example, a group interested in family life problems might take a forum program if it could find nothing in its own area of interest. However, such a shift rarely occurs. If the program it has heard goes off the air, the group simply disbands.

A number of well-known programs have gone off network schedules or off the air altogether, taking their groups with them. The *Maddy Band Lessons*, the *NBC Home Symphony*, *Wings for the Martins*, *You and Your Government*, and a number of other programs have been restricted in coverage or have disappeared altogether, and their numerous groups have perished. Occasionally a questionnaire sent in to the Study bewailed the loss of such a series. Frequently, of course, a program exhausts its usefulness, and should be terminated. Few will survive a long term of years. We must therefore count on programs dying, and on a considerable mortality among groups as a result of their doing so.

Similarly, local stations will occasionally cease to broadcast certain network programs, particularly those in evening hours or at daytime periods which meet the competition of baseball games or the determined wishes of a profitable sponsor. *America's Town Meeting of the Air* has an unusually full network coverage, greater, in fact, than that of most commercial programs. Yet a few stations to which the program is offered do not take it, and about half a dozen use transcriptions and present delayed broadcasts at different hours from that at which the program comes direct. Some stations have broadcast this particular program direct during some seasons, and in succeeding ones have offered it by transcription at another hour. Such changes may result in the disbanding of groups; although, as we have seen in the case of Syracuse, a transcription may attract considerable group activity.

In the case of other programs, the hour of the broadcast may be changed, and this often causes a falling off in the number of groups following it. Several leaders in their reports, and some members in their questionnaires, complained of changes in time as deterrents to group activity. Groups in New Jersey, Oklahoma, and California all made such complaints.

"Very inconvenient," writes a New Jersey leader of the 11:45 hour for *Homemakers' Forum* which had forced her group to listen to the broadcasts separately, and come together at a later time for discussion. (The time has since been changed, but still not to the satisfaction of many groups.)

"Friday at 3:15 is too late," reports an Oklahoma leader, explaining that many mothers had to collect their offspring from school at about that time and could not attend.

"The one o'clock hour was too early for the majority of housewives," writes another leader from San Francisco with respect to the *Family Life Series* in that state. She adds: "Changes of time are alway confusing. Someone forgets the hour, particularly when dates and time are made or set so far ahead."

In all the above cases the groups persisted, but under handicaps. In other cases they did not even persist. *Homemakers' Forum* officials in their annual reports have attributed the loss of groups—sometimes running into the hundreds—to changes of time, and there is no doubt that both their program and a number of others have suffered in this fashion.

What's Wrong with the Program?

The quality of the program is an element over which groups have no direct control, and program quality is a factor in the disintegration of some listening units. Undoubtedly the failure to devise suitable programs is a much greater factor in preventing the formation of groups—I shall comment on that point in a later chapter. Just now we are concerned with the extent to which poor or unsuitable programs kill off listening groups.

When one considers the quality of most of the programs to which groups listen, one is surprised that the programs themselves do not discourage listeners more frequently than they apparently do. To be sure, in relatively few cases are the offerings downright poor. On the other hand, a number of them are merely respectable. Too frequently they are straight talks not well adapted to large audiences, or dialogues that are somewhat stiff or unnatural. Again, the entire broadcast may show signs of hasty preparation and lack of rehearsal. In too few cases has much attention been given by program makers to the possible needs of the groups. To be sure, some broadcasts are well planned and excellently produced, and pretty well adapted to group use. However, they are in the minority.

The rather low level of program quality—as judged from the standpoint of group needs—does not greatly discourage most listening bodies. They accept what is given them with what I sometimes regarded as a remarkable placidity. In some instances their patience can be explained by the fact that the broadcasts are sponsored by a local P.T.A. or club, and the listeners wish to be loyal to their organization. In other cases group listeners quite clearly are convinced that any form of education must be fairly dull. They are interested in getting information, and this is given to them. They are grateful for the gift, and would not be so presumptuous as to demand entertainment or stirring interest along with teaching!

However, not all groups are placid under what they consider to be poor program quality. Leaders sometimes complained seriously of the character of the programs they heard.

"We feel that the broadcasts this year do not offer much sulutions for the avradge family problems," an Oklahoma leader writes, with spelling and grammar as given.

A California leader complains: "The programs seem to be more appropriate for a teacher or student of family relations than something the busy housewife can absorb and put into life experience."

A Middle Western group for which a leader filled out a questionnaire stopped using radio entirely chiefly because of the character of the program. "We tried the program last year," writes the leader, "meeting in the evening after having heard the broadcast (individually) in the afternoon. We received sufficient material for a lesson but really derived no help from the broadcast as it was so poorly delivered and came to us transcribed—not really interesting or appealing. This year we are using a course of study as submitted by the *Parents' Magazine,* so you see our group was really quite disappointed in the radio broadcasts."

These complaints were exceptional, yet there were a number of others somewhat milder in character which emphasized the importance of the radio offering itself to the success and even the existence of the group. Makers of programs who want groups cannot afford to ignore these signs of disappointment or downright dissatisfaction. The longer a group follows a program, the more discriminating and exacting it is likely to become.

We Are Betrayed by What Is False Within

However, the groups reporting to the Study were more likely to show a consciousness of faults within themselves or their communities than of weaknesses in the programs they heard. Many leaders or would-be leaders reported apathy or inexperience among those who were or might be members, or complained of other activities which crowded out group listening.

"The group for some reason have lost interest," says a CCC commander in Iowa.

"About twenty people attended the first two meetings," says a Pennsylvania librarian who was doubtful of the future, "but there were only twelve at the fourth meeting. . . . It seems rather difficult to stir up public discussion in this town."

Other reports more vague in character are of a similar type.

"For several reasons we have not succeeded in getting a group on a successful basis this year," writes a cautious Connecticut Y.M.C.A. secretary. "If replies could be sent in three weeks later we might be ready to work with you."

An educational director in a Nebraska Jewish Community Center states: "Our local listening group did not meet with any success this year. In its stead we have organized a current affairs discussion group."

Even groups that were active had their difficulties with members.

"Some come only three or four times a year," one leader complains from Oklahoma, "others one or two times. Only a few come every time."

"Not enough interested parents," writes another leader.

"It is difficult sometimes to get people away from their firesides and radios, down to the village hall," says a representative who was more energetic. "We find it necessary to do a lot of telephoning to get people interested."

"There are so many activities in a town of this size," asserts a California leader, "that one must constantly be choosing what to do."

There were some remarks by leaders to the effect that better study aids would promote the health of groups, and there were a few mild complaints about gossip usurping the place of serious talk. However, these comments were relatively inconsequential. The chief weakness of which leaders and members alike were aware has in fact not yet been mentioned in this chapter, although it has been referred to in previous ones. This is a lack of good leadership.

Leaders Wanted

Some if not all of the groups mentioned above as ailing or failing because of the character of their members were probably suffering chiefly because those in charge of them were in one way or another incompetent. An able, trained leader can make something of almost any group. Excellent bodies of radio listeners have been developed in rural districts having no town facilities at all, and in towns with no more than six hundred inhabitants. Of course, much depends on membership too, but that is less important. However, a leader who might do well with one group of people—say, students or professional men and women—might do badly with another, such as farmers, laborers, or housewives with little schooling or income. Naturally, many persons seeking to organize groups are unaware of their own shortcomings as organizers or leaders, or may have poor judgment in picking those who will have responsibility for the activities of the groups.

"It is hard to get good leaders," says a leader himself in referring to the fact that the leadership of meetings does not rotate in his group.

"The group is not meeting at present," reports an official of a community organization. "Interest was not sustained. Leader became involved otherwise."

"The women are difficult to keep to the subject when there is not a trained leader," admits the representative of a P.T.A. local in California.

"Our group only met once a month," says a Michigan woman.

"There were only about five that took any real action or attended. There was no one actually to lead or organize."

Even groups that have shown astonishing bursts of life suffer from a lack of wise direction.

One leader had asked for 75 questionnaires to fill out. He was sent 25, since the larger number was not desired from one group, and he eventually returned 7. Clearly his group was beginning to wabble. It had thrived for a time, staging a special discussion of its own which was broadcast by a local station, in addition to listening to *America's Town Meeting of the Air*. The trouble was indicated in the questionnaire—the undertaking had been too ambitious. "Our local speakers are not always up to snuff," the leader said in the space provided for general comment at the end of the questionnaire, "and it is increasingly difficult to get good ones. We are a bit confused as to the next step."

Probably a wise leader would never have attempted so much with the resources at hand; and had he done so, he would have seen no difficulty in revising his program to eliminate the broadcasting by the group, or would have planned to attempt it only occasionally. Many groups that lead a vigorous life would be incapable of maintaining the role of broadcasters in addition to performing those of listeners and private discussants.

Personal visits to groups showed that many could have profited by better leadership. Clearly those in charge of thriving organizations needed to be alert, and often were seeking means of improving their meetings. Some were uneasy about the future. "Do other groups have difficulty in maintaining interest?" one young man asked me rather wistfully. The most successful were those who had developed several persons—or half a dozen—able to lead.

It is interesting to note that 224 members of groups felt that they had a "spark plug" without whom the group would not be likely to go on, while 168 felt that their group could continue even if the most obvious leader in it dropped out. Of similar interest is the fact, revealed by the leaders' reports, that 35 of the 105 groups represented by these questionnaires always had the same leader for their discussions. If these groups were representative, then a notable proportion of all groups depend on one person, and might disintegrate if he should cease to function. As we have seen, groups actually did fail because their leaders "became involved otherwise."

Some Live Long

The mortality of groups was approached in several other ways than those indicated above. Individual members were asked if they belonged to other groups, or had previously belonged to them, and if any such groups had failed, and why they failed. The supposition was that if groups were constantly dissolving, a number of their members might form new connections—that perhaps from 20 to 33⅓ percent of all members might have had previous group experience. This did not prove to be the case. Only 34 of 484 members, or about 7 percent, reported previous membership in other radio listening groups, and only one member reported that a group to which he had formerly belonged had died. His explanation of its demise was brief: "No interest."

What then? Do radio listening groups live a long time? A means was provided for getting some evidence on this point. Leaders were asked to give the dates of the founding of their groups, and only one failed to do so. The year of founding and the number of groups reported for each year were as follows:

Year	Number of Groups
1926	1
1932	2
1934	1
1935	2
1936	4
1937	10
1938	30
1939	41
1940	14
Total	105

Naturally, this evidence is no more than tentative. It may be argued that the older groups were more likely to come to the notice of the Study, or that the proportions in the above table are accidental, and would not hold if 1,000 groups instead of 105 were canvassed. Naturally such arguments cannot be met. They may be valid, although I incline to believe that the 105 groups heard from were fairly representative. Certainly they heard a variety of programs, were located in various parts of the country, and were not selected because of any knowledge of their age. In any case, we find more than half of the groups that returned questionnaires claiming an age of at least two years, while at least three-quarters were more than a year old (some of the 1939 groups were of less than that age).

It is therefore clear that some radio listening groups live a long time, and it is probable that a large number of them go from season to season, and can hardly be called "fly-by-nights." Naturally, their membership may change from year to year, although there is evidence to show that even such change is not great, for of 5,353 members whose leaders reported on their length of membership, 3,324 had been active for more than a year. We must, I think, grant somewhat more stability and endurance to groups and those who compose them than most casual observers in the past have been prepared to attribute.

Factors in Vitality

What then shall we say about the diseases and mortality of groups in so far as we have evidence?

The factors making for health or decline are at least fairly clear. The withdrawal of the program from the air, whether locally or altogether, the quality of the program, the time at which it is broadcast, and frequent changes in time, are factors beyond the control of groups; yet such factors will or may be decisive. Within the group the character of the membership and, particularly, the quality of leadership, will be the chief influences on group life.

Naturally, other factors may play a part. Organization, the character of the outside aid that may be received, the congeniality of members—all these may be important. In so far as help from without the group may develop confidence and leadership—and it can do much toward encouraging both—such assistance may be crucial. In the prevention of mortality, this factor offers in my opinion the greatest hope for the future health of groups. I shall come back to it at a later point.

However, much remains to be done in program making and scheduling which may also promote group vitality. Some of the programs to which groups listen or fail to listen could be greatly improved. They suffer still from the educator's belief that schoolroom or straight lecture techniques can be applied to radio education, and from the broadcaster's willingness to take what the educator gives him. They suffer also from station network relationships, and from the casualness with which some radio officials will shift about an educational program at the behest of a commercial sponsor.

Naturally, many of the defects from which groups suffer innocently are the result of the fact that radio practice is still in the making. The tendency has been strong on the part of broadcasters toward better programs, fewer changes of time, and a greater sense of obligation to the public interest. Doubtless we shall have in the future a radio industry that increasingly does more thoughtful and constructive work in the field of education; and in producing more instructive programs of high quality, with firmer schedules, also creates a better environment for listening groups.

IX • WHAT PROFIT?

THE WOMAN sits on the sofa in the painfully clean, somewhat overcrowded living room of her little home.

"Oh, I think it's been one of the greatest opportunities I've had," she says. "You see, I didn't go far in school—only a year beyond grammar grades. But I wanted to do right by my children, and I didn't know how. Then I learned about this program the university has. I listened, and I found that professors and doctors talked to us. They took up the same situations most of us have to face every day. And often—why they'd do the same things I'd have done; only I'd have worried about it. It was a fine thing to get their ideas. Then I found we could get books, too, and pamphlets. I wouldn't have known what to read, but this way I could find out. I've been telling all my neighbors about it."

She was a member of a group.

Later, in a large city, the man who talked about groups was not a member of one. "It's a way of keeping the political and social heritage which we've had, and are in danger of losing," he said. "Through the radio, we can bring thousands of meetings in America into action. They can hear men and women who can speak on a given question with authority. They can discuss what they hear. They will take a new interest in what affects the nation. Many of them will read. Most of them will think. I'd like to see ten thousand local forums preparing their members to be citizens."

He was the director of one of the great public discussion programs on the air, a program which encourages the formation of radio listening groups, and serves them.

In another city another man, in charge of a radio station,

also spoke about groups. "Yes, we want them," he said. "All intelligent station managers do, I think. For one thing—and this is a purely selfish reason—they are a tangible evidence of what radio can do in the public interest. Of course we don't think only or even chiefly of that, but it's important. Yes, we want listening groups."

These statements could be multiplied many times. Thousands of group members would speak enthusiastically of groups, hundreds of educators and radio executives. Perhaps we have seen already a number of reasons why they would do so. Yet thus far we have stuck fairly close to facts and have done a minimum of analyzing and reflecting. The time has come to review the facts and ask their meaning.

What the Educators Get

Let us still keep close to the tangibles. Let us ask first what are the advantages for an educator in the listening groups which he or others acting with him may encourage to listen to a program which he devises.

There are of course the usual advantages of radio—its capacity to assemble a large audience, to present these listeners with distinguished speakers or with programs of high artistic quality, musical or dramatic. These advantages multiply the numbers of people who can be informed or taught, and they make possible a skilled and careful preparation on the part of the educator. However, we may take such benefits for granted when we think of group listening in contrast with individual listening. When we contrast group work with nonradio education, we must on the other hand bear the particular resources of radio in mind.

For the moment let us compare group work with other radio listening. It has definite characteristics which recommend it to those trying to teach by air.

First of all, the radio educator wants an audience, and groups insure one. It may be said that he has an audience even if he has no groups, and to an extent this is true. Indeed, he may even check his listeners by telephone or other surveys, and discover their approximate number. He may also get many letters from those who hear the broadcasts which he presents. However, he can have little confidence in the attentiveness or steadiness of the listeners who may or may not hear his program, and who have no definite obligation with respect to it. All he knows is that a certain number of radio sets are tuned in. Many are pouring their message into empty rooms; or fighting against a family conversation on other matters; or against an absorption on the part of the listeners in household chores, a novel or a textbook, a quarrel between children in an adjoining room, or the imperative summons of a doorbell or telephone.

But if the educator has from 200 to 3,000 listening groups, he knows that from 2,000 to 50,000 persons are almost certainly giving his broadcasts their serious attention, and that a number of them are making what they hear a point of departure for further reading, work, or study.

He knows too, if the results of the Study of Listening Groups have validity, that his group listeners are talking with others about the programs and often urging them to listen. Groups thus act as promoters for an educator's program.

Finally, the educator achieves a two-way communication with his audience which is far superior to what he would enjoy if he had no groups. Naturally, he could learn through "fan mail" the opinions of some listeners, and could even correspond with individuals if he wished to do so. However, most fan mail offers only a limited basis for dealing with the audience. The individual listener who writes about the program is sometimes helpful, but often is not. For example, he may be an irregular listener, or one whom the educator does not care to reach, or one about whom the educator can learn relatively little (although recent studies make an analysis of fan mail possible within broad limits) unless he spends a considerable amount of time and energy.

In contrast, the listening group represents regular listeners definitely interested in the educator's work, who can speak to him out of thought and experience. The educator can of course consider other listeners also, but the group gives him easy access to one important type of hearer. Listening bodies are ready to make suggestions about the time of the broadcast, the character and quality of it, the kind of study aids most desired, and various other matters. In many ways they resemble an audience which the educator can see personally—he can indeed send representatives to listen with groups and to observe them—and if encouraged to express themselves, they will frequently be even more articulate than such an audience. The organized listening unit, if utilized by a program maker, pretty thoroughly does away with the supposed greatest disadvantage of radio—that of being a one-way form of communication.

In the second place, groups can be used by the educator to justify his program to radio authorities or to educational administrators; or they can be made a means of attracting speakers or participants. Listening bodies can also help the educator to demand more of those who perform for him.

All such advantages are well illustrated in the Oklahoma program, *Family Life Radio Forum.* Dr. Alice Sowers, the director, can point to her groups as an evidence of widespread interest in her work. It is safe to say that owing to the groups she was able to establish the Family Life Institute, has convinced the university authorities of the vitality of the program, has got more radio stations throughout the state to carry it, has gathered a larger number of individual listeners, and, as a result of groups already established, a larger number of groups. For the knowledge that bodies of radio listeners exist stimulates the formation of new ones. Moreover, Dr. Sowers has undoubtedly had an easier time in getting speakers to appear on the program because of the serious popular support that her groups indicate, and she has been able to influence her speakers to prepare more carefully. She gives a sheet of instructions to each expert before he appears, and it is safe to say that participants usually pay more attention to these instructions because they know of the listening bodies which the program attracts, and are inclined to feel the importance of their roles more than they would were they appearing before a radio audience which to their knowledge was composed only of scattered individuals.

The groups also encourage and justify the preparation of listener aids, and thus give the educator an opportunity to intensify his work.

Finally, the existence of organized listeners may enable an educator to get funds or the equivalent with which to improve program quality. A station or network will be likely to spend more on a program which has support such as listening units represent. As a program evokes such public response, the radio authorities will at least be willing to provide improved facilities for production if these are necessary. Naturally the educator profits as his resources for producing his program are increased, and the radio network or station and the general public are both likely to profit too.

How the Broadcaster Profits

The broadcaster profits indirectly in all the ways in which the educator profits. He is committed to a reasonable amount of education by air; often he is an eager promoter of it, seeking educators who will carry out his purpose, and striving to improve the quality of the educational offerings he sponsors. It is greatly to his advantage, as one serving "public interest, convenience, and necessity," to be able to point to numerous bodies of listeners organized to hear one of his programs. He, like the educator, wishes to know what listeners like and want and need. He desires a two-way communication with them. As the representative of one network stated in a recent conference, he welcomes groups of people who by their acts in organizing say: "We are going to listen," for too frequently he hears from individuals and organizations who have objections to programs and say, "We are not going to listen."

However, the development of groups on any considerable scale serves the broadcaster in a more tangible way. They tend to increase and maintain his audience. A program like the *Family Life Series* in California with 1,631 groups is certainly one which most broadcasters would prefer to other educational offerings without known groups. The group activity is also usually marked by the presence of a large number of individual listeners who cannot listen with others but do listen by themselves. Dr. Sowers in Oklahoma has taken a creative step in offering aids to such persons, thus establishing them as registered listeners. As we have seen, *America's Town Meeting of the Air* has a considerable body of "registered" individuals following its broadcasts. Any program with listening groups is likely to have a large fringe of serious individual listeners, and can if it wishes register and serve a considerable number of these.

Moreover, all groups and most registered listeners act as agents for spreading the fame of the programs they hear. It will be recalled that 404 of 484 group members testified that they talked with others about the programs they heard, while 353 urged others to listen. If the same percentage of all probable group listeners in the United States act similarly, we might think of from 200,000 to 250,000 persons acting as unpaid agents for educational programs with groups. Registered listeners, it will be remembered, were quite as active. Of 423 who filled out questionnaires, 371 reported talking with other people about the program for which they were registered, and 326 urged such persons to listen to this program. In addition, 311 registered listeners and 172 group members reported that as a result of their group or solitary activities they tuned in more frequently on the station broadcasting the program for which they were registered, or which they heard with a group, while 313 registered listeners and 254 group members said that their activities had made them seek other educational programs on the radio.

Thus the groups and the registered listeners seem in many ways to assist in maintaining the radio station's audience. Naturally, their influence should not be exaggerated. In many cases it is relatively small. Yet it constitutes a partial answer to the broadcaster's assertion: "If I put on an educational program, I lose listeners for my next sponsored program." In any case, the groups and registered listeners will definitely cut down his losses. With careful program planning and more extensive utilization of the advantages which groups and registered listeners offer, radio officials may find the groups and the serious-minded individuals a notable asset from a purely selfish point of view.

Balance Sheet for Listeners

What a listener gains or stands to gain from groups and registered listening is more important than the gains of either the educator or broadcaster. Both by law and by reason, service to the listener is the chief purpose of radio, to which commercial profit and educational theory alike are subordinate.

This service may be thought of as comprising two principal types.

There is that rendered to the great majority of the population, including our sixty odd millions with an eighth-grade education or less, and with small desire for more learning than they already have. As yet we have done relatively little to educate the less schooled and less thoughtful Americans, chiefly because our traditional pedagogical activities have been designed, consciously or unconsciously, for persons with at least a full high-school education, or with a mental keenness which has provided them with an equivalent of that amount of schooling. We have entertained our eighty million radio listeners pretty well; we have educated them badly or not at all.

In contrast, for the ten or fifteen percent of listeners who are fairly well educated or mentally aggressive, we have done better. Most of our educational programs have been devised for them, and certainly we should give them as much in the future as we have given them in the past. This minority vitalizes American society, and is entitled to radio service, as is the majority whom we shall have to learn how to serve more effectively.

Let us bear in mind, then, that serving the listeners is the chief obligation of educators and broadcasters, and that were there a conflict between the interests of listeners and the latter two, the listeners should come first. But let us realize that there is unlikely to be much of a conflict. Educators and broadcasters tend to prosper as listeners are prospered, for service to listeners is the chief end of their efforts.

What do groups and programs for registered listeners do for the American radio audience? The actual and potential benefits are many, and worth noting specifically. I shall put them in outline form for purposes of clarity.

1. When he listens with a group or as a registered listener, an individual gives a care and attention to what he hears that he does not otherwise give, and gets a decided benefit from this finer quality of listening.

   Of a student group one leader writes: "Many students do not enjoy the privilege of listening to radios when they like. This is because they live in centers where radios are ruled out at certain times, or are for other reasons unavailable. The radio forum as

we conduct it gives these youngsters a chance to hear an A-1
program undisturbed and unforbidden."

"When our group listens, we *listen,*" declares a woman representative of a P.T.A. body.

"They've learned to listen as well," says a Y.M.C.A. secretary of
group members.

Such testimony tallies with what an observer sees when he visits
a group. There is never any conversation or other distracting
activity during a broadcast.

2. The broadcast brings a listener material of a quality he cannot
   easily get elsewhere. Leaders and members were particularly appreciative of the kind of speaker presented by radio.

3. From his group activity the individual gets an opportunity to
   hear the opinions of his neighbors, thus enjoying the benefit of a
   local as well, perhaps, as a national discussion of the subject.

4. He gets practice in discussion. Of 484 members filling out questionnaires, 230 reported that they always joined in group discussion, 152 that they did so frequently, 55 that they did so seldom,
   while 47 did not answer.

   Again, 299, or about 65 percent, felt that they discussed more
   freely at the time of answering the questionnaire than when they
   had first joined the group.

5. The individual gets a social pleasure from his group. Of 484
   members, 367 valued their group membership because of its
   social aspects, and 60 felt that seeing friends or making acquaintance was a chief reason for belonging to a listening body. Again,
   216 more felt that social pleasure was definitely a reason, although not a highly important one.

6. Through his group, the individual is stimulated to read and act
   in addition to listening carefully and discussing. We have already
   seen the extent of such activity. The registered listener also receives a comparable stimulus.

7. It is clear that through group listening an alertness and aliveness
   of outlook is promoted. Testimony on this point has already
   been given.

8. The radio group is a comparatively inexpensive way of getting
   information and education; and it is also a convenient means of
   doing so. Many leaders and members noted both these advantages.

   "It doesn't cost anything! I don't have to worry about finances!"
   says one of the former.

   "We feel that at no time, at less trouble, could we obtain a
   lecturer or speaker comparable to what we get with our forum
   broadcasts," writes another.

9. Although really deriving from discussion, open-mindedness may be classed as a particular advantage gained by listeners from group associations. The following comments on this point are typical of extensive testimony:

"We have learned to be tolerant of opposing points of view."

"The broadcasts have done much to promote clearer thinking, articulate expression, and tolerance of opposing points of view."

"Have also learned to discuss as well as to argue."

"It has liberalized my attitude toward my employees."

10. Finally, as previously brought out, group work offers advantages which perhaps a majority of group members could not get otherwise. Statistics have been given on this point, but a few personal statements may help to make the feeling of group leaders and members clearer.

"We haven't any local organization outside this group at all!"

Of other organizations: "Too formal, too large for real active discussion."

"A town of our size could not make speakers of the type [offered by radio] available. We could go to Philadelphia, but this saves time."

"There were no classes conveniently available in our county."

"Local discussion on church affairs only. . . . The quality of radio is higher."

"The women's club is fine, but does not study children."

As can clearly be seen from the figures and quotations given just above and earlier in the book, group members and registered listeners were keenly aware of many of the advantages which groups brought to them. It may be asked, however, how they felt with respect to the entire experience as compared with other types of education. Some evidence has already been given on this point, but certain additional testimony should be considered here.

For example, members were asked as to whether or not they felt that their groups did for them pretty much what a class, club, or nonradio group would do. A total of 379 of 484 said that they did; 58 answered negatively, and 47 did not answer at all. This result indicates that most of those in listening groups believe that they are receiving a fairly intensive educational experience.

As already pointed out elsewhere, members also assessed their group progress under the headings "Made little or no gain," which only 12 members checked; "Gained somewhat," checked by 305; and "Have made great gains," checked by 139. In addition, they checked possibilities designed to compare their impressions of what they learned in the radio group with what they thought they would have learned in a nonradio group. Of the 484 concerned, 63 made no answer to this question; 105 thought they had learned about the same as would have been the case had they engaged in nonradio group work of some kind; 298 thought they had learned more; and 18 thought they had learned less.

Thus on the whole group members had a fairly high opinion of the quality of group activity. Many were enthusiastic about it, writing "Much more!" in their answers to the last question referred to. Yet there were a number who did not rate their radio group experience very high as education.

"I believe I gained more listening alone," says one.

"With a good leader, and a sincere effort to learn, I feel I could have gained as much by a nonradio group," says another.

A third member gives his opinion that classes "are far more valuable than listening to the radio. But the Town Hall radio programs are a good supplement."

"I gain more by watching the speaker," states a woman in a P.T.A. group. "The radio during the day makes me nervous."

Still others felt that the group work was thin. "The charge of superficiality is always present in my mind," declares one leader. Others recognize limitations, but feel that for many persons it would be learning a little by radio, or learning nothing in other ways. "People will listen to the radio," says one group representative. "They *don't* always *read*." "Our women are busy and they will listen for fifteen minutes, then talk and discuss a little when they would not get it in any other way," says another.

What Rating as Education?

Such comments bring us to a question which has been discussed with respect to radio education in general. How does it compare with various other forms of education? Clearly the answer has a definite bearing on the value of listening group activities.

Quite as clearly, I think, it can be perceived that for most group members what they get is not comparable with intensive class work in regular educational institutions. For a small minority it may be, but this minority must be fired by a passion for learning, and make extensive use of opportunities for outside reading. Perhaps from 15 to 25 percent of the members may be aggressive enough to get as much or more from radio listening groups as from a high-school or college course. Personally, I doubt if more do.

Yet if the work of the groups is to be compared with that of most adult education classes, both the testimony we have and the results of extensive personal observation tend to show that the listening groups can hold their own. They stimulate, they broaden, they give definite information which is pretty fully assimilated, and they get a good amount of special work accomplished as a result of what they incite members to do. Adult education is education for people who often do not have the time or inclination to carry on intensive work. Its methods cannot be designed to produce the amount of effort which those of many school or college courses require. It can well be satisfied if it shapes constructive habits and starts a mild amount of activity. All this the radio listening groups do.

In general, they command more industry on the part of their members than nonradio discussion groups. I am satisfied of this both from evidence as to what is actually done, and from personal observation. Few other agencies in adult education bring people together so frequently—let us remember that 59 of 105 groups reporting met weekly for a considerable portion of the year, or for all of it. Few other agencies combine listening with a lecture or prepared talk, extensive discussion by group members, and a reasonable amount of study. Indeed, if the persistence from year to year of many groups is remembered, and the influence on outlook, reading, thinking, and action is considered, there is a good case to be made for groups having an effect upon many members which is comparable, although in a different form, with that of classes doing intensive work.

The two types of education are different. Formal education works in fields that are usually more specific; groups tend to deal with those aspects of life which may influence citizenship. Making better parents and better voters is highly important work—work, incidentally, that the schools have neglected in the past, and are seeking to cover more fully today.

When we think of group work in the terms of broader culture we perceive its importance to the listener. The value of what has been done is I believe considerable; the value of what may be done is much greater.

And in considering such facts and possibilities we must recognize that in a democracy the listening groups do more than serve broadcasters, educators, and individual listeners. They serve society as a whole. They promote leadership, they make for a better-informed electorate, they encourage the habit of discussion on which a democracy must rest, and they break down prejudices and increase tolerance. Moreover, they already do these things on a greater scale, in proportion to the effort expended, than any nonradio agency can hope to rival; and they hold the possibility of exerting much more power than they have already manifested. Radio and its groups may be as important to the democratic process as the railroad and motor car are to transportation. They can be multiplied and immeasurably improved. If they are, they may become the instruments of a constructive and significant revolution, a revolution not only in educational techniques, but also in important social habits.

X ⋅ ADVICE FOR THE AMBITIOUS

"HOW do you go about starting a listening group?"
That is a question which, with many variations, often ended the hunt for an active body of listeners to a radio program. Letter after letter came to the Study, in response to our inquiries, saying: "We do not have any groups, but we should like to learn what you discover about them, for we have thought of undertaking some such activity." Educators, station officials, and representatives of various organizations wrote to this effect. Often the writer, seeking information, found himself seated in a Y.M.C.A., Y.M.H.A., or a club house, answering to the best of his ability eager questions about groups. It seems illogical to close this study without attempting to deal with some of the many queries made in the course of a year by interested persons of varied types.

Since whatever is suggested in the next few pages will be read with the greatest interest by persons who may want to use it, I shall first set forth what I have to say in a simple outline. The chief divisions of this can then be discussed briefly.

How to Develop and Maintain Radio Listening Groups

1. First, for the encouragement or actual starting of listening groups, a suitable radio program must exist or be created.
2. Broadcasters or educators who want to encourage listeners to form groups to listen to a radio program which they are producing should:
    (a) publicize the possibility of starting groups
    (b) offer assistance in the way of advice by letter or in person.
3. Broadcasters or educators seeking to encourage groups to listen to such a program should if possible also offer a service in the form of printed or mimeographed aids.
4. In connection with any large-scale program for developing groups—say from 150 to 3,000 of them—provision should be

made for training group leaders. This is not so formidable an undertaking as it may appear to be.

5. In starting an individual group, those in charge should remember that groups can be of various sizes, each having different needs and a different character. Such types are:
   (a) the intimate group—from 5 to 10 members
   (b) study or discussion clubs—from 15 to 35 members
   (c) forums—50 members or more.

6. It should be borne in mind that:
   (a) a small group may be better in certain circumstances than a large one
   (b) the larger the group, the more care and the more resources are usually required for its success.

7. A group can be started:
   (a) by an individual or several individuals without any support from local agencies or institutions
   (b) under the auspices of a local agency—library, club, church, etc.
   (c) by the joint action of a number of individuals and agencies —as when a group is used as a community forum.

8. Every group needs:
   (a) a capable leader, for conducting meetings
   (b) a "spark plug"—who may also be the leader, or may be merely a member
   (c) members interested in the subject of the program the group hears, and willing and competent to participate in group discussion.

9. If a group needs vitalizing, the following means of renewing or sustaining interest may be employed:
   (a) publicity, to draw in new active members, and awaken the interest of existing ones—news stories, notices on bulletin boards, announcements at local meetings
   (b) reminders—postal cards and telephone calls
   (c) outside speakers of reputation
   (d) visitors with special interest in and knowledge of the subject of a broadcast, not to be invited as formal speakers, but to join in discussion
   (e) as objective a study of the group as can be made, with the idea that it may be attempting too much, or in other ways not working in the most efficient manner. Action can then be taken to correct errors of policy or procedure.

From the above outline it can be seen that I have taken into account two kinds of persons interested in group listening. These are (1) the broadcasters and educators, and (2) group leaders and members. The broadcasters and educators have charge of radio programs and wish to encourage or promote the formation of groups. They have no desire to become members of particular bodies of listeners. The leaders and members have just that desire. Each wants a group of his own to which he can belong and from which he can derive personal profit. He is little concerned about how many other groups exist, except as their existence may help him to get better programs and better study aids. The distinction between these two kinds of persons should be kept in mind, as I shall deal sometimes with the problems of one, and again with those of the other.

Let us now turn to the matter of the outline, and consider each item of it in some detail. That of the radio program to be used for group listening comes first.

First Get a Program

Both experience and common sense plainly indicate that before there can be radio listening groups there must be a radio program suitable for group listening. A broadcaster or educator can attract no groups unless he can broadcast such a program. No individual or local organization planning to do group work can proceed to act unless he can find a suitable radio offering which a listening body can use effectively.

What kind of a program is needed?

No satisfactory offhand answer can be given to this question, for it is probably the most important that can be asked with respect to group listening, just as it is vital in connection with listening in general. However, certain characteristics can be listed as essential to a program suitable for groups.

In the first place, the program should deal with a subject of wide general interest. While there might be exceptions, groups are not likely to be attracted unless the subject is of such a type. We have seen that listening bodies have been most numerous for programs dealing with family life problems, and with problems of political and social importance. Two fields of greater interest would be difficult to find. But we have seen too that musical programs of various types, and programs dealing with education, gardening, drama, news, and vocational questions have also attracted groups. Still other subjects might, and undoubtedly will, prove suitable for group listening. So long as the interest is sufficient, the subject does not matter.

The subject should be so developed as to stir and hold the imagination of its listeners. In other words, subject is not enough; within the chosen area of interest a program idea should be created that will command attention. Fewer people may be interested in gardening than in the welfare of their own children, and interest in the former is unlikely to be as deeply serious as interest in the latter. But if a program on gardening were developed with high skill and dramatic appeal, it might clearly have more chance to attract groups than a program on the problems of child-rearing which might be dull and uninspired.

The program should give information that is useful and authentic. This attribute scarcely needs discussion. Groups in general gather with the expectation of being informed.

The program should suggest opportunities for discussion. This characteristic has proved to be highly valuable for group work. Some groups have indeed met steadily merely to listen— as for example to an opera. Yet even these tend to remain more vigorous when they discuss what they have heard, and in the case of programs dealing with family problems or political questions, discussion undoubtedly awakens the interest and enthusiasm of group members. Where conflicting points of view can be presented and argument invited, the group profits.

The program should create the desire to know more about the subject it deals with. All radio education must in its nature be stimulative rather than complete, and the capacity to arouse curiosity and an eagerness to know more is valuable in any educational program. If the program is presented imaginatively, and if it stimulates discussion, it probably also stimulates the desire to read or practice or otherwise act after the broadcast has been heard.

Questions of Quality

The above characteristics might be those of any good educational program. They are, nevertheless, the characteristics of a program suitable for group listening, and those attempting to plan such a program, or those hunting for a program already being broadcast, would do well to note these simple attributes.

Needless to say, the question of getting a satisfactory program is much more complicated than this brief description would indicate. We should bear in mind that in all broadcasting the program is the central fact with which we deal. The benefit to most listeners is limited by it; if it is mediocre, the profit to them is meager; if it is glorious, most of them will share its glory.

With listening groups the quality of the program is bound to have a tremendous effect. We may in one case have a dull program which never gets any groups—we can never know the number of groups that poor programs have prevented from being born! Or perhaps with the aid of an organization like a P.T.A., groups are formed and maintained by a sense of duty for a program which in reality does not deserve their attention. Or perhaps, again, we have a well-conceived and brilliantly executed program which attracts and keeps groups. In all cases, program quality influences the existence of groups—their numbers and size; and it also influences, often profoundly, the kind of work they do.

Unfortunately we still have far to go in America in creating programs which are really satisfactory for group work. A few such programs exist, but the majority of those now being heard by listening bodies are merely acceptable. In general, our programs suffer because they are planned, written, or produced without distinction. Thus the chief need is for raising artistic standards. However, there is also the possibility of modifying programs that are of fine quality so that, without loss of value for individual listeners, they will serve groups more effectively. To carry out such a possibility an increasing amount of study on the part of program makers will be necessary. Such study has already been undertaken, both in Great Britain and the United States. In March, 1939, the British Broadcasting Corporation published a two volume mimeographed report by its educational research director, Mr. R. J. Silvey, on listener reactions to four programs broadcast by the B.B.C. in 1938. Both individual listeners and listening groups supplied material on which the report was based.[1] In April, 1939, the Advisory Service of Town Hall, Inc., furnished a preliminary digest of returns from a questionnaire of fifty questions sent to its groups with the object of getting their opinions on aspects of *America's Town Meeting of the Air* and the study aids furnished in connection with the program.

Such investigations will enable those planning programs to act with a much greater knowledge of what listeners think and feel. As a result, programs should have a much more direct relationship to groups than they have had in the past, and, it is to be hoped, a higher quality as programs.

Making the Program Known

Granting that a suitable program for group listening exists, a highly important consideration in the mind of the broadcaster or educator in charge of it who may want groups is to make the program known to such persons or organizations as are likely to be interested in establishing bodies of listeners.

I have already pointed out that once the idea of forming groups is publicized, a number of persons or agencies are likely to initiate them. Also, some groups may already have sprung up and will be heartened by encouragement from program headquarters, even if this encouragement is accompanied by little in the way of an offer of tangible assistance. However, some assistance must be offered, and the more the better. The minimum would be an invitation to register with program headquarters and receive a schedule of future broadcasts, together with advice by letter (and possibly by personal visit) with respect to the problems of group organization and procedure.

Let us say that those in charge of a program are prepared to invite the participation of groups and to offer schedules and personal advice. How should they go about advising "prospects" of their interest in groups, and of the assistance they are prepared to give?

Clearly the easiest and one of the best ways is announcement

[1] *Listeners and the Autumn (1938) Talks.* A Report on an Enquiry into Listeners' Opinions on "Men Talking," "Everyman and the Crisis," "Class," "The Mediterranean" (Listener Research Section, British Broadcasting Corporation, March, 1939).

by radio. Such announcement should be made during the program (if it is already on the air), and also at other intervals in the radio schedule. It should state briefly the advantages of group listening, set forth the willingness of the program sponsors to coöperate with groups, and tell interested persons where to address inquiries.

Wherever practicable, radio announcements can also be fortified by news stories. For newspapers friendly to educational radio, the mere fact that a station or an educator with an informative program is launching a service to groups is good material for an article. Furthermore, a news story printed some time in advance of a program's first broadcast can be effective in getting groups organized to follow the program from its beginning.

A follow-up story telling of the establishing of the first group, and later stories reporting on the number of groups functioning after a month has elapsed are good journalism from the point of view of a city editor or a radio editor friendly to broadcasting activities. News articles about groups are of course likely to be read by some persons not reached by radio announcements. More important, such publicity will make group work seem more interesting and more important to those who have already heard by air of the possible formation of groups and of the services available to them.

Circulars briefly describing the program, the possibilities of group work, and the assistance which program headquarters will give are also important means of stimulating group activity. Naturally, circulars should be sent to a carefully selected list of individuals and organizations. Like news stories, they can be used considerably in advance of the first broadcast of a new program.

In particular there are literally scores of organizations already existing, all of which should at least be informed directly of the possible relationship of groups to the series likely to attract them. Which of these innumerable agencies active in forming groups should be circularized will naturally depend upon the particular program offered. If it deals with public questions, then agencies such as school systems, private schools, libraries, colleges, social work agencies like Y.M.C.A.'s, churches, professional clubs, and many other types of institutions may profitably be informed about plans for group work. In the case of a family life program, women's organizations certainly should be circularized. If the broadcasts are presented during evening hours, the field of agencies interested would be broader. As we have seen, even for a daytime program, the *Family Life Radio Forum* of Oklahoma has appealed to a number of men's associations. However, common sense will indicate what kind of a list should be used for circularization in the case of any given program.

Circularization can be cut down in many cases by establishing personal contact with organizations. If the educator or broadcaster will approach representatives of a state congress of P.T.A.'s, for example, explain to them his hopes and purposes, and show what service he is prepared to offer, the agency so approached may approve the program and undertake to sponsor or promote it. I have already pointed out that a great variety of organizations do sponsor particular programs, and urge their local branches to listen as units. The possibilities for personal contact work—local, regional, or even national—need not be canvassed in detail. Obviously such possibilities are varied and numerous.

Servicing the Groups

Let us assume, then, that those in charge of a program suitable for listening group activity have publicized the fact that they want groups, and are prepared to give assistance to any which will organize. Let us further assume that this assistance in the beginning consists merely of advice by letter and the furnishing of schedules giving a brief description of the series, with the titles and dates of each broadcast.

Clearly, even if no more than this is done, some person or persons will have to do advance promotion work, and later will have to correspond with groups, make up schedules and see that these are produced in quantity. All such work except possibly the correspondence could be handled by an educator himself or by a radio station official, or could even be undertaken by the representative of an organization enthusiastic about the program. Perhaps, as in the case of the California P.T.A. program, *Family Life Series,* an organization is fully responsible for the program—

is, indeed, the educator. Of course, for any extensive correspondence with groups, a stenographer would have to be available in addition to the person directing promotion and correspondence.

Those promoting the program and corresponding with groups would naturally work during whatever part of the year the program was on the air, and, if it were being launched for the first time, during a preliminary period of some months preceding its inauguration. Usually an educational program runs for a "season"—*America's Town Meeting of the Air* is broadcast for twenty-six weeks. However, in the case of this program, group activity is so extensive that an advisory staff of three executives and a number of assistants works throughout the year.

If program officials are seriously interested in group listening, they usually render more service to groups than that outlined above, and of course it is desirable that they should if that be practicable. Let us look at possible features of such service, and at the demands which these will probably make on a staff in charge of ministering to groups.

Study Aids Almost a "Must"

Few programs followed by a large number of groups fail to offer printed or mimeographed aids. We have already discussed these in some detail in Chapter VII, and it will not be necessary to go over the ground again. However, in my opinion groups should receive at least the following materials:

(a) a handbook for leaders
(b) directions for members as to the technique of discussion
(c) questions on the subject of each broadcast
(d) a list of books, pamphlets, magazine articles, etc., bearing on the subject of each broadcast
(e) some kind of background material on the subject of each broadcast—such as a brief presentation of the chief possible points of view, with excerpts from books or magazines.

For groups of inexperienced persons, such materials are almost a "must" if successful work is to be expected. It will be remembered that most groups returning questionnaires to the Study received a comparable amount of material, and that their leaders greatly appreciated having it. A service comprising the five items listed above can be provided with little effort and expense. Items (a) and (b), the handbooks for leaders and members, can be prepared in quantity, and usually need to be sent out to a group only once. When a large number of groups follow a program from year to year, the two handbooks can be kept in stock.

Indeed, there is no reason why these little guides should not now be produced on a national scale and made available cheaply to educators and broadcasters having only a modest number of groups. Such booklets are as usable by one group as by another, regardless of the subject matter of the broadcasts being heard. The Federal Office of Education, the National Association of Broadcasters, or Town Hall, Inc. could prepare standard guides and print them in ten thousand lots. The costs could thus be cut to five or ten cents a copy. I make the suggestion that this be done. The simplifying and promoting of discussion is a matter of concern to all Americans, and guides for accomplishing this should be made available at a nominal price.

The other three items on the minimum list given above require constant attention, and of course must be related to the particular program for which group aids are being offered. They will change from week to week. Naturally, they must be prepared each time by someone thoroughly familiar with the program, presumably by the person in charge of relations with groups, or by an assistant. The amount of work required to formulate them is, however, not formidable—the entire job can be done in from a day to three days, according to the kind of service being rendered. Nor are such materials expensive in the form commonly used—that of two to four mimeographed or multigraphed sheets.

In fact, even if the promotion of groups is undertaken by program headquarters all the work—the furnishing of advance schedules for the year or half year, correspondence with group leaders, and the preparing and dispatching of aids—can be maintained on a modest scale by a director and one assistant, both working full time. This type of service has in fact actually been rendered by two persons for almost two hundred groups. However, in the instance I have in mind only a single state was covered by the program, and the aids were not as full as they should have been. With more groups, a greater territory, and a more comprehensive service, a larger staff would naturally be required.

There is a distinct possibility of doing more with the "background material"—that is, the presentation of chief points of view, and the quotation of excerpts from authorities—than has yet been attempted. Some members who filled out questionnaires for the Study complained of the fragmentary nature of this information. To be sure, the book lists were designed to suggest supplementary reading, but many protested that they made little use of such lists. "I wish," said one gentleman, "that the actual reading material could be sent me. I will not take time to go to the library and look up material, but I will read such material if it is placed in my hands." His suggestion merits study. The National Policy Committee, in its Public Affairs Pamphlets, shows how a great deal of information, admirably written, can be put into compact and inexpensive form.

Let the Groups Pay

Those in charge of programs may protest: "This is all very fine, but we have no money for financing any extensive amount of printed aids."

True; but why should the broadcaster or educator finance them? If these aids can be produced cheaply, groups themselves will pay for what they get. The cost is not great, and people usually appreciate something for which they are charged a moderate amount more than they do something which is provided free. Town Hall has charged for its extensive service to groups, and has found plenty of units able and willing to bear the expense.

A minimum service could, indeed, be offered for as little as $2.50 a group for a season of half a year, the equivalent of ten cents apiece to a membership of twenty-five. A $5.00, $7.00, or $9.00 fee might be charged for more extensive aids—these prices have been set by Town Hall, groups being offered a choice of various combinations on a list of a dozen services, the cost to the group varying with the amount of service desired.

Leader Training

An additional service which has been offered in only a few cases, but which might easily be developed for programs followed by a large number of groups (or even as the joint enterprise of several programs broadcast by the same network or from the same station) is leader training.

I have already dwelt upon the importance of this activity, and we need not discuss its significance again. However, a word may be said as to its practicability. Adequate leader training requires able directors, quarters where meetings can be held, and a reasonable amount of time. The accepted plan, and the most feasible where adults engaged in group work as an avocation only are concerned, is to hold short institutes, usually lasting several days. From a few of these, any individual with natural leadership abilities can get a great amount of profit. For a single agency to undertake such institutes is quite possible, but may be difficult. For several to combine in such an activity would appear to be easier.

In the case of certain types of group work, a number of organizations profit. For example, in New York and New Jersey many radio stations, Y.M.C.A.'s, churches, CCC camps, school systems and clubs all have an interest in Town Meeting groups, and directly or indirectly profit from what they do. The whole activity of group listening is an aspect of the democratic process of preparing men and women for citizenship, and therefore all society is benefited where groups are doing efficient work. Can not and should not a number of these organizations unite to hold several leader institutes in the course of each year? Certainly there can be found among them ample resources for the work.

The groups themselves can send experienced leaders to talk with inexperienced ones, and to give demonstrations. Schools, colleges, Y.M.C.A.'s and clubs can furnish instructors. Educators in charge of the program often have the ability and experience to outline a brief but efficient program, and to help direct it. Meeting places and an institute clerical staff may well be supplied by the various organizations interested. Where a radio council exists, as for example in the Rocky Mountain area near Denver, the council could take general charge of leader training. Where such councils do not already exist, a temporary organization could be devised.

Indeed, the failure to provide leader training would seem to lie rather in a lack of recognition of its importance than in any insuperable difficulties in the way of carrying it off successfully. Perhaps as groups grow and prove their value more widely, leadership training will follow automatically. Meanwhile, if it could be established the quality of group work should improve.

The Group Itself

We have now discussed the chief factors involved in establishing programs for groups, in encouraging group activity by publicity and contact work, and in providing groups with adequate service. Within the framework of these efforts the individual listening unit would function. Let us look at this unit. How can it be launched and kept in vigorous health?

Anyone who faces the problem of starting a group should remember what has previously been pointed out—that there are three general types: the very small or intimate group; the medium-sized discussion club; and the forum. Membership for these would be, respectively, from 5 to 10 persons; from 15 to 35; and 50 or more.

Clearly an ambitious father of a group should look over the membership material he has to draw upon, the character of the community in which he lives, and other relevant factors, and aim at a membership suited to the situation he faces. A small neighborhood on the outskirts of a town of 5,000 people is not likely to have the resources for a successful forum; a town of 25,000 persons or more may have such resources. But even a few people can combine to make a successful small group. Seven persons, for example, may do intensive and profitable work. Perhaps a unit of from fifteen to twenty-five persons is the ideal for discussion as contrasted with speech-making. Would-be organizers or leaders should keep these facts in mind, and not attempt more than they can successfully perform. Size is not in itself a merit.

Of course the community forum, if ably managed and conducted, may have a sparkle and dramatic lift of its own. If it is wanted and there are local leaders and speakers to make it successful, then by all means it should be planned for. The important thing to remember is that any group functions in a definite environment and is limited in its possibilities by that environment. Those who launch the group should not seek to make it something which its surroundings do not permit or encourage.

A group can be organized by a number of different persons or agencies. One enthusiastic individual can and often has launched a successful listening body. Several friends working together may do so. A local institution, such as a church, a Y.M.C.A., or a college, may initiate a group. Finally, a number of agencies, representing an entire community, may pool their enthusiasm and resources. The Study has records of numerous groups founded in each of these different ways.

What is the procedure? As I have previously pointed out, many leaders reported in effect: "I asked a number of my friends to come and listen to the program; we discussed it, and enjoyed our experience so much that we organized and continued to meet regularly." Conversation among friends has brought about similar results. Colleges or Y.M.C.A.'s approve the idea of a group, make announcements and post notices, provide quarters and a leader, and the group comes into being. Where a community group is established—of the forum type—an adult education council or a voluntary committee usually takes charge of the work necessary for the launching of the listening body.

What a Group Needs

If a group of any kind is to be successful, it needs certain resources. It must have some person willing and competent to act as leader; preferably it should have several such. Naturally, leaders can be developed if the group is enthusiastic and studies the aids sent out by program headquarters—granting, again, that these aids tell how to organize a listening body and how to conduct meetings. Usually they do. Usually, too, some of those willing to form a group have had experience in presiding at meetings, and often in holding conferences, an activity closer to the work of smaller groups.

Assuming that a leader exists or can be developed, the group should also have a "spark plug." The leader himself may be a spark plug—often he is. On the other hand, he may be able so far as the conduct of meetings is concerned yet lack the qualities which will permit him to act as a kind of human ignition agent for the listening body in general. Such an agent should have an enthusiastic conviction about the importance of group activity. He should be able to impart his enthusiasm to others. He should have a knack of making his associates work, of bringing out their capabilities. He should have a sharp eye for defects in policy or practice. He should have the industry to undertake himself, or the ability to delegate to others, the work that may be necessary for the making of arrangements for meetings, the sending out of postal cards as reminders, or the making of telephone calls to round up lethargic members. Often the person who performs such services is not the leader at all. He may be an unobtrusive member who nevertheless gets around, plugs up gaps, sees that the wheels turn smoothly. He is likely to be a quiet and astute man of all work rather than a brilliant speaker.

A leader should ask himself if he is a good spark plug; if he thinks he is not, he should try to find someone among his members who is. A group with a good spark plug seldom dies from internal weaknesses!

Every group should have members really interested in meetings and discussions, and willing to participate. Small groups for this reason often thrive by building up their membership from the friends of the founders, these friends in turn bringing in persons whom they know and judge to be suitable for membership. Thus the group is hand-picked for character and interest. At the same time, the general tendency seems to be to search for variety of occupations and background, along with social homogeneity.

However, many groups consciously seek variety without too much regard for social relationships. Those discussing public questions are particularly eager to get "a cross-section of the community." This cross-section is a selected one, yet it is based on seriousness of interest and mental keenness rather than on friendship. Thus bankers, mechanics, teachers, nurses, laborers, unemployed, business men, clerks and doctors often will be found meeting together. They have a certain bond in their common interest in discussion, but they represent a variety of experience and opinion. Such variety provokes argument and brings out different points of view. It tends to make for lively meetings, and at the same time to encourage tolerance.

The larger group or forum needs all the elements discussed above, but it needs something more: able speakers and members who can and will ask questions and talk effectively from the floor. Thus it must have greater resources in public speaking talent than the two smaller types. If its founders are not sure that these will be available, they had better think twice before launching their group forum.

The larger group also needs publicity in its earlier stages, and usually throughout its existence. It cannot hope to gather from 50 to 250 people regularly without letting the community know about its doings, and without sustaining the interest of the community in them. In other words, it is putting on a show, and the show must constantly be advertised. Also, as a consequence of its character, such a group cannot be selective as to its membership. It must open its doors to the public and be prepared to have a shifting audience. Naturally it should be sure of a core of faithful members who will come pretty regularly and help to keep the meetings vital. There is usually no difficulty in getting publicity for or community interest in a group during its earlier stages; the difficulty in keeping people aware of its activities and eager to share in them is likely to come after the group has been established—often when it believes that it has solved its problems!

When the Group Falters

What can be done to keep a group successful? Often it starts its life with enthusiasm, and seems headed for continuous success. Then it begins to languish, and leaders and members wonder what they can do to recapture the first fine enthusiasm and enjoyment.

A number of things can be done. If the group is small, someone should use the mails or the telephone to be sure that a good proportion of the membership comes out. Often members intend to come, but forget that a meeting is to be held, or need prodding. If the group is large, and depends upon publicity, someone should see that notices, announcements, and news stories advertise the character of each meeting. Aids to groups often give instructions as to getting publicity. Usually these, with a little imagination and industry, are sufficient for success. Indeed, many groups include persons willing and able to see that the group receives its share of public attention.

The local library will usually coöperate by posting notices. The mere fact of a meeting and the data on subject and speakers will generally win a small news story in the local paper. Community organizations will make announcements if the group has won its place in the town or city where it meets.

Changes in program often revitalize a group. If there has been a tendency toward general discussion in a group of moderate size, an occasional shift to a debate may stir up new interest. If formal debates have been held, an attempt at a more informal meeting will often succeed.

Groups will sometimes profit, too, by inviting outside speakers to visit them. It will be remembered that of the 105 groups covered by the leaders' questionnaires, 10 reported always having outside speakers at their meetings, while 36 stated that they had them occasionally. Such speakers tend to attract visitors and keep up the interest of regular members. In general, groups have used them little. It will also be recalled that the questionnaire for leaders contained inquiries as to whether teachers, judges, office holders, industrialists, etc., had assisted the groups by speaking to them, and that the response showed such persons to have visited listening bodies but rarely. Again, as we have already seen, only 35 groups reported coöperation from their libraries. In most cases where no help from outside speakers or agencies was reported, the groups were probably to blame for the lack of it. Small units may not need the help which they might get from the outside. The more pretentious ones often do. I believe they could get it in abundance if they sought it. The resources at their command seem in most cases scarcely to have been scratched.

Many groups could increase their vitality without the necessity of appealing to community leaders. Often, indeed, the leaders are not the "best bets" for stimulating member interest or the interest of the public. Is a program dealing with the problems of youth being heard by a group? Why not bring to this particular meeting several young persons, and draw them into the discussion following the broadcast? This was actually done by one group which I visited. Similarly, labor officials could be invited to a meeting dealing with labor problems, while P.T.A. units could stage (as some do) occasional meetings at which fathers and children appeared, as well as mothers. In college towns, student debaters can be asked on occasion to lead a discussion. Representatives of various vocations who can furnish interesting testimony on certain subjects may be invited to be present at the appropriate times. Bankers, engineers, doctors, army officers, shippers and others would have been appropriate guests at certain times for groups hearing Town Meeting programs last year.

Most groups have not been alert in utilizing local resources of the kind suggested. A few have done so. Often the success of a listening group may depend upon the wise use of such resources.

Finally, any group which feels that its work is falling off will do well to take stock of its policies and performances. Is it attempting too much for its resources? Is it attempting too little? Are its procedures a bit boring? I have seen groups in action which were suffering from one or another of such errors, and were unconscious of them. A realistic consideration of what is being done and what can be done might result in changes which would be constructive. To understand their own problems, groups or leaders of groups would do well to visit other listening bodies, observe what these do, and consider whether or not they themselves can learn from what they have observed. Apparently there have been few exchange visits among groups to date. There should be many more of them.

Building and maintaining a good service to groups is not easy. Building and fostering a group is an arduous if exciting and enjoyable activity. Too much care cannot be given to either undertaking. The work in either case will sometimes thrive with a relatively modest amount of labor, but usually it will not. Those who want to promote groups or share in a particular group's activity should alike be ready to do some thinking, planning, and vigorous participating. They should expect obstacles and difficulties. Listening groups cannot draw their vitality from a loud speaker; often they must struggle for it.

Still, there is a reward for diligence and patience in this field of organized listening. A well-organized, well-serviced, and vigorous group is a delight to a visitor, a member, or a leader. It is doing an important work for its neighborhood or community, and usually the group members are having a considerable amount of fun as they improve themselves and make their indirect contribution to a better citizenship.

APPENDIX

"HOW DID YOU BECOME INTERESTED IN THE LISTENING GROUP TO WHICH YOU NOW BELONG?" (REPLIES OF MEMBERS TO QUESTION 8)

Cause for Member's Interest	Number of Times Checked
Talk about listening groups at a meeting of a club or group to which the member belonged	211
Talk with a friend or relative about a group	166
Hearing a radio announcement about listening groups	61
Ways not specified on the list of possibilities	61
Notices in newspapers	46
Talk with a librarian	16
No answer	25
Total (for the 457 members who answered the question)	586

"CHIEF PURPOSES" OF MEMBERS IN JOINING LISTENING GROUPS (QUESTION 9 OF QUESTIONNAIRE FOR MEMBERS)

Member's Chief Purpose in Joining Group	Number of Times Checked
To get information on public questions	90
To coöperate with your club or other organization, etc.	84
To get a better education	59
To share in group discussion	57
To enjoy meeting people	25

"CHIEF PURPOSES" OF MEMBERS IN JOINING LISTENING GROUPS (QUESTION 9
OF QUESTIONNAIRE FOR MEMBERS)
(Continued)

Member's Chief Purpose in Joining Group	*Number of Times Checked*
To improve your skill in some art, craft, or hobby	8
To see what education by radio could do	7
Other purposes than those listed	5
To save time by "keeping up" through radio instead of by reading, etc.	1
Total (of members giving a chief purpose)	336

SECONDARY PURPOSES OF MEMBERS IN JOINING LISTENING GROUPS
(QUESTION 9 OF QUESTIONNAIRE FOR MEMBERS)

"Other Purposes" Involved in Joining Group	*Number of Times Checked*
To share in group discussion	220
To enjoy meeting with people you liked	169
To get a better education	159
To get information on public questions	150
To save yourself time by "keeping up" through radio instead of by reading, etc.	119
To coöperate with your club, etc.	97
To see what education by radio could do	95
To get information more cheaply than you could in other ways	48
To improve your skill in some art, craft, or hobby	7
Other purpose—not specifically listed	20
Total	1,084

PART II

▪ ▪ ▪

GREAT BRITAIN

With Some Reference to
Other European Countries

▪ ▪ ▪

BY W. E. WILLIAMS

PREFACE

THE DOMAIN of group listening in Europe is extensive but homogeneous. In sifting the evidence on which this Report is based it becomes clear that there exist only two kinds of organized group listening. The first kind is that undertaken voluntarily by a group of people with the avowed intention of giving the broadcast some further collective consideration: the kind which proclaims its purpose by being so often called a wireless discussion group rather than a listening group. The other kind is that which is included in the compulsory curriculum of an educational institution. In quantity the second kind far exceeds the first. Many European countries which have no use for wireless discussion groups have a well-established service of school broadcasts; yet this second type of group listening is much easier to evaluate than the first. A third variety of group listening is that based on the scarcity of radio sets among poor or primitive communities: an "accidental" use which can be dismissed from this inquiry.

European practice in school broadcasting will occupy less attention in this Report than group listening of the other sort. Its scope is well defined in every way, and it is free from that problem of group leadership which so persistently besets the organization of wireless discussion groups. School broadcasting has a limited function: it serves as an interesting variation of the routine of class-teaching; it provides additional material beyond what the teacher can provide; it introduces interesting personalities into the classroom; but at its very best it is a supplement and not a substitute. It belongs to that category of auxiliaries which includes the film, the school journey and all those other agreeable variations of the time-table which the modern school so sensibly employs. Moreover, in school broadcasting the emphasis is on instructional value rather than discussion value. The travel talk introduced into the geography lesson serves as an illustration and not as a stimulus to debate. The readings from literature are of the same nature; so are the lives of famous scientists or episodes from history. For the most part, the school broadcast is an extension and an illumination of the day-to-day time-table.

It is a subsidiary service, valuable and agreeable, but never a substitute for the careful concentrated instruction given by the teacher—nor has any wider function been claimed for it. Listening groups, on the other hand, aim at an educational self-sufficiency which we must examine in some detail. In some European countries an attempt has been made to use groups as a primary educational service, but nowhere has such a possibility been envisaged for school broadcasts. This distinction, then, accounts for the emphasis which will be placed in this Report on the scope and limitations of voluntary group listening by adults, rather than on the system of school broadcasts which have become universally accepted as having a well-defined place in school timetables.

The preparation of this Report has been made possible largely by the ready coöperation of broadcasting authorities in Great Britain and several other European countries. They have put at my disposal many internal reports and tables of figures which have helped to present a picture of group listening in Europe; and they have been very willing to discuss matters of policy in the most candid way. Apart from this valuable assistance, I have drawn upon other sources of information and opinion, viz: (a) my experience as a member and an officer of the Central Committee for Group Listening in Great Britain during the period of development of this new educational medium; (b) my visits to innumerable listening groups, to training courses for group leaders, and to conferences on group listening arranged by adult education organizations; (c) investigations made, by interview and by questionnaire, for the purposes of this Report.

One way and another the testimony of more than 300 witnesses has been considered before a line of this Report was written down; and although it would be vain to expect their unanimous endorsement of all that is set down in these pages, my effort has been to prepare a Report that could command general agreement among those most familiar with group listening on this side of the Atlantic. Of the many individuals who have helped me to arrive at my conclusions I most particularly thank Mr. Roger Wilson and Mr. N. G. Luker of the B.B.C.; Mr. G. W. Gibson and Mr. W. R. Reid of the Central Committee for Group Listening; Dr. Ygve Hugo of the Swedish Radiojanst; Mr. A. R. Burrows of the Union Internationale de Radiodiffusion, Geneva; and Mr. Paul Bensinger of the Société Suisse du Radiodiffusion, Bern.

Although none of them must be regarded as having a responsibility for the opinions expressed in the Report, their willingness to assist my inquiries was one of the pleasantest features of the whole investigation.

W. E. WILLIAMS

British Institute of Adult Education,
29, Tavistock Square,
London, W.C.1.
April, 1940

I ‧ SOME FUNDAMENTAL
CONSIDERATIONS

THE FIRST doubt which comes into the mind of anyone who is sceptical about group listening is this: Has it merit in itself or has it merely the merit of ingenious adaptation? Salesmen of vacuum cleaners often overcome the resistance of a dubious housewife by luring her into some ridiculous ambush: they convince her not by emphasizing the primary purpose of their machine, but by mentioning in an aside that the vacuum is useful for shampooing her dog. We must watch out for a similar snare in assessing the value of group listening. Is it really valuable in itself? Is it an authentic new educational method? Or is it a freak application of the primary purpose of broadcasting? Does it owe its origin and development merely to the fact that somebody said: "Here is a powerful new mechanical instrument; let us see what tricks we can play with it"? It is from this cautious and agnostic angle that group listening should be approached.

The Slow Progress of Group Listening

The dimensions of group listening in Europe suggest that, whatever its merits may be, they have not yet attracted audiences of any impressive size. In Great Britain, for example, the constituency of group listening has never yet exceeded a total of 20,000 in one year, despite the fact that time and money and organization have been more freely spent during the last twelve years on this type of adult education than on most others. In other European countries, save in Czecho-Slovakia, the "reaction" has been no greater. Sweden has not equalled this total of 20,000; nor has Finland or Norway. Russia has published the usual astronomical figures to prove her citizens' insatiable appetite for culture, but no more credence can be put upon such figures than upon any other statistics proceeding from such unverifiable sources. Ironically enough, it is Germany which might claim the leadership in the mere volume of group listening. For in the Third Reich, and nowhere else, is it possible to bring the entire nation to a standstill to hear a talk on the whole duty of a good citizen. Discussion of the "talk" on such occasions, however, appears to be limited to the mass repetition of adulatory noises.

It is not, of course, reasonable to compute the group listening audience as a fraction of the entire population: it must be reckoned in relation to the numbers engaged in similar cultural activities. But even on this reckoning, it is still true to say that group listening has made less headway in the European democracies than any other comparable educational activity. In Great Britain the aggregate of group listeners has never been one-tenth of the student membership of organized adult education. As an independent activity it represents a very small section indeed; its numbers are largely due to the fact that group listening has been taken up as a subsidiary activity by the bodies engaged in more advanced and continuous kinds of adult education. These considerations, which are correspondingly true of all the European democracies, must be borne in mind in approaching our first question: Is group listening valuable in itself or is it a novelty which cannot of its nature live up to the claims which were forecast by its more sanguine pioneers?

The Balance Sheet of Group Listening

The immediate advantages and disadvantages of group listening are not difficult to discern. On the credit side there can be set down, first of all, the fact that the method of group listening permits a far higher standard of authoritative exposition than any other mode of education. A dozen artisans in a small town can, by this means and this means alone, enjoy the satisfaction of listening to the nation's highest authorities on a wide variety of topics. Although that experience is too often undervalued by the critics of group listening, it must be admitted that it has limitations. Of these, the outstanding one is the fact of the speaker's inaccessibility; not only is his personality etherealized, but there is imposed upon him a rigidity of presentation. However sympathetic his nature, he is, in a sense, talking in the air. He is cut off from all opportunity of noting his audience's response to his argument—an opportunity which, even in such minor matters as frowns of doubt or disagreement, is one of the most valuable guides an expositor can have. Closely related to this limitation is the further disability imposed upon the discussion which group listening is designed to instigate. The speaker is absent from the debate he has provoked; and no vicarious representation of his views by the group leader can effectively make up for that absence. Anyone who has been present at a large number of listening groups cannot fail to be struck by the fact that more often than not the discussion is *ersatz*.

Many thoughtful members of listening groups dwell on this profit-and-loss aspect of the method; and on the whole they come to the conclusion that the eminence of the speaker is more than offset by his subsequent absence from the debate. Group listening can summon rare spirits from the vasty deep, but it cannot give them that embodiment which is the nature of the effective teacher. Here, indeed, is the fundamental problem of group listening. It can carry the lesson up to a critical point with unrivalled authority; but at that critical point the guidance of that authority is withdrawn and replaced by a leadership which is inevitably of lesser quality in every respect. And even if it were not lesser in quality the fact that it is different is an equally substantial disadvantage. The process which occurs when the talk finishes and the group begins its discussion under its leader is the risky process of changing horses in midstream; and as the change is often from a pedigree charger to a nondescript nag it is not surprising that listening groups so often fail to get across. At every stage of the investigation which this Report records, the fact has emerged that group leadership is the clue to the popularity or the failure of group listening. There are, of course, other factors in the reckoning—the choice of subject, the manner of presentation, the nature of the meeting place, the quality of reception; but even when these problems have been solved there remains the crucial factor of group leadership.

It is a healthy sign that this fundamental disability of group listening should nowadays be recognized—and nowhere more plainly than among those responsible for the organization of group listening. In the early days of its development in Great Britain group listening suffered from the excessive claims made for it by some of its sponsors. No one would today be so reckless as to regard group listening as a primary educational method; and since it has come to be assessed as the secondary method which it undoubtedly is, there has been a corresponding advance in the discovery and identification of the uses to which it can best be put.

The Theory of "Special Uses"

In theory, at least, two such uses stand out. The first is the listening group as an approach to a more exacting and continuous course of adult education. In this sense a listening group should be able to break up fallow or stubborn ground which later on can be more intensively cultivated by other systems of adult education. There is, in fact, little evidence to show that group listening serves this purpose, nor even that it attempts to do so. The percentage of people who seek more elaborate adult education after membership in a listening group is negligible. The group listening audience is largely composed of two contrasting types: people who are already members of some sort of adult educational course, and people who are not likely to be. The first category are so interested in educational activities that they welcome any such auxiliary service as broadcasting. The second category find that group listening just about satisfies an educational interest which is both elementary and intermittent. The first category are the type from which some excellent listening groups have been made; but the second category seem to represent the average level of group listening. But as a "feed" for adult education, group listening has not fulfilled the expectations foretold by some of its pioneers. Whether those expectations were sound and probable is another matter.

The second special utility which group listening was expected to reveal was its value to rural districts where more formal facilities are difficult to organize. No European country has been rich enough or willing enough to afford a really adequate system of rural adult education. Such factors as the travelling expenses of tutors and the thin "plantations" of interest still make it difficult to cover the ground in country districts. Yet here again group listening has not reached the objectives which theory prescribes for it; and listening groups, save in Czecho-Slovakia and to some extent in Norway, do not flourish particularly in the villages. Effective groups depend, as this Report will illustrate over and over again, on leadership, and good leaders are harder to find in the country than in the towns. The leaders of social and cultural life in the villages are a strictly limited number—a schoolmaster or two, a clergyman, and so on—and the claims on their time are always heavy. Some of them manage to find time to lead a listening group; but the fact remains that the comparative unavailability of leadership personnel retards the development of group listening in rural areas.

Mention may be made of a third special adaptability which group listening might be expected to possess: its particular value for such isolated communities as prisons, sanatoria and the like. In Great Britain there have existed a handful of such groups, but nowhere else in Europe does there appear to have been any attempt to organize groups of this kind. In some ways it should be easier to establish group listening in places where the population has time on its hands, but there are often peculiar obstacles to be overcome. A Prison Governor may not fancy the idea of giving a convict that authority which group leadership involves; long-term convalescents are afflicted with psychological conditions which can undermine and blow sky-high a cultural interest which they adopted with enthusiasm the day before. And in both cases suitable leaders are often hard to find. Yet to judge by the few experiments which have been made in Great Britain, there are in this special field possibilities which would reward persevering efforts.

Group Listening's Limited Appeal

On the whole it seems that group listening has not managed to mark out for itself territories which were unoccupied or uncolonized before. Except for a small salient here and there, its sphere of interest lies inside rather than outside the established frontiers of adult education. In most countries the administrators of group listening prefer that it should be so; in Great Britain especially the responsible authorities take the view that a group listening movement as such would be undesirable, and that their purpose should be to "build the listening group into the existing educational fabric." It is an auxiliary or supplementary service rather than a pioneering activity. Although this view may be disappointing to some exponents of educational broadcasting, it is one which is supported by considerations familiar enough to everyone who is experienced in adult education movements. Education in any direct or frontal form has never attracted big battalions of interest, and it probably never will. That does not mean that cultural interests generally are not enlarging their appeal and increasing their numbers of adherents. But it means that "educational" interest in European countries is a pervasive rather than a deliberate concern. It manifests itself, for instance, in slowly increasing standards of film-going and concert-going; in the striking growth of "serious" reading; in the membership of political and co-operative organizations; in the demand for informative periodicals. Broadcasting, too, represents one of the most important of these informal educational agencies. That type of broadcasting which we call group listening may only attract thousands; but the feature programmes and the general talks programme attract tens of thousands and hundreds of thousands. The intellectual interests of the community are active enough and avid enough—but they refuse to be canalized in listening groups. What is more, there is evidence to show that the talks arranged specially for listening groups are heard and appreciated by a far bigger audience outside the group organization.

The Discussion Fetish

The modest growth of group listening, despite the care lavished upon it in Great Britain and some Scandinavian countries, is due neither to a general lack of intellectual interests nor to a failure to appreciate broadcasting as a feeder of such interests. It is more likely due to a widespread indifference to the wireless-discussion method; and for that indifference it must be admitted there is good ground. By calling the tribe together to sit round such a totem pole as a wireless set you do not inevitably ensure a valuable exchange of opinion and knowledge. In some of the most experienced and reliable quarters of adult education there is a reaction against that glorification of discussion which has become a fetish of adult education. There is a kind of discussion which awards prematurely the privilege of an opinion, and which is an impediment rather than an approach to knowledge. The indispensable necessities for effective discussion are that it should be led and directed by someone of incontestable authority, and that it should be based on some preliminary knowledge of the subject. If these conditions are absent, discussion degenerates into a vain beating of the air; it can even have the baneful effect of loosening the toe hold which a thoughtful student is beginning to get on the subject, and thus disorganizing his entire interest and attention. Corroborative evidence of this danger inherent in group listening can be found in other kinds of study. In the course of this investigation a group of college students complained that they had been disappointed at the results of a study-circle which their professor had urged them to conduct in their own way, in their own time, and without his guidance. They found, again and again, that their discussion lacked authority, and therefore kept losing its way in a No Man's Land of assumption and speculation. There is nothing which more rapidly—and rightly—discourages the thoughtful citizen than the discovery that a group discussion (of any kind) is not being conducted by a dependable authority. He soon sees that he is getting nowhere, that he is even wasting the time which he might be spending on finding his way alone.

These considerations are set down here not in order to denounce discussion as an educational method, but to emphasize the absolute conditions in which discussion can be effective and to suggest that these conditions are at least as often absent from the wireless listening groups as they are from some other elementary forms of adult education. Many thousands of thoughtful people, discovering this liability of the listening groups, do not repeat their first acquaintance. For many more thousands, having encountered this sterile kind of discussion in other places—in the club, the pub, the railway carriage—arrive at the reasonable conclusion that they require at their stage knowledge rather than argument. These are the unnumbered hosts who prefer fireside listening to group listening; and the fact that their interest is not limited to the momentary experience of hearing a talk or a feature is evidenced by such matters as (a) the extent to which they follow up suggestions for private reading; (b) the volume of sensible letters which they write to the broadcasting stations. The fireside listener is not less interested and not less intelligent because he doubts or dislikes the experience of group listening.

The "Lebensraum" of Group Listening

The comparative indifference to group listening in Europe must not be mistaken for an indifference to broadcasting as an educational medium, much less to education as such. France, for example, is so unconvinced of the virtues of group listening as to show no disposition even to give it a small experimental trial; yet some of the French programmes supply a very considerable "dosage" (as they call it) of "serious" talks intended for studious people at their own firesides. The same is even more true of Sweden, Denmark, Holland, and Switzerland. Their programmes not only carry a high percentage of instructive talks, but they include a large number of popular self-contained foreign language courses. Yet despite this impressive amount of educational broadcasting, none of them is convinced of the desirability of group listening for discussion purposes, and two of them have not even felt disposed to attempt a group listening organization. It would certainly be most inaccurate to suggest that the existence of a group listening organization is any kind of proof of a community's belief in and desire for broadcast educational provision.

Some broadcasting officials from whom an opinion on this point was sought concealed their doubts about group listening by the diplomatic excuse that their systems, unlike the B.B.C., could not afford to finance and officer a group listening organization. Others, in Holland, Denmark, and Switzerland, suggested that adult education is so plentiful, accessible, and specialized in their countries as to leave no room for a further providing agency. French witnesses observed with disarming candour that their countrymen were such practised and incorrigible debaters anyway, that there was no need to equip them with the sort of discussion training which group listening offers to the dumb and diffident English; and one of them suggested that if Great Britain were equipped for the *terrasse* life of France, she would find it possible to train her citizens in the art of conversation without any cost to her public funds. The statement is not so preposterous and flippant as it may seem.

In the European democracies—in the countries, that is to say, where freedom of choice persists—this paradox appears: group listening can get nowhere unless it is sustained and fostered by the providing bodies of adult education. Yet where those bodies are powerful and well established they appear to leave little *lebensraum* for such a newcomer as group listening. Group listening cannot take root unless it is cultivated by the adult educational organizations; and if those organizations are flourishing, they do not find it worth their while to give time and space to the cultivation of what they regard as a minor crop. To this paradox there must also be related the fact that in no European country has there been forthcoming a really hearty and vigorous support of group listening by the adult education organizations. Sometimes they have given it mild support in so far as they perceived in it a possible source of recruitment to their own membership. Sometimes they have participated against their inclination lest other bodies should capture these hypothetical recruits; and to the extent that group listening has failed to round up such recruits their interest has dwindled to a polite tolerance.

Such considerations alone, however, do not account for the tepid attitude they have adopted to this potential new agency. On the whole, they have seen the fundamental weaknesses of the group listening method, and they have concluded, rightly on balance, that some of these weaknesses are inherent in group listening and that others can only be removed at considerable financial cost. Of these, the principal is the inadequacy of the group leader—a factor which will persist as long as these leaders are willing but untrained amateurs. The adult education organizations on the whole believe that group leaders will not be effective until they are trained and remunerated; and since that remuneration would be a charge on the comparatively small public funds now available for adult education as a whole, they feel that a better case must be made for group listening than the one so far apparent. And what clinches their dubiety is the feeling that if money is to be spent on paying more tutors, it should be paid for tutors who will undertake a fuller range of tuition than is afforded by the leadership of listening groups. From all points, therefore, authoritative opinion within adult education cannot bring itself to regard group listening as a primary and positive educational agency. So far, that opinion assesses group listening as an instrument of occasional and limited value. One very experienced witness, commenting on the way group listening in Great Britain has grown so slowly, despite the time and money so freely expended on its development, put a general opinion in these racy terms:

> Group listening, unlike other types of adult education, has to be continually bolstered up. Without tireless injections of publicity, bribes, bursaries for Group Leaders to attend training-courses, the patient keeps collapsing. The Education Officer who acts as doctor drives round the countryside at great expense imploring patients to keep alive and acting as midwife at unwilling births. This, I know, is also true of some other modern educational movements—but is it healthy? And was it true of any of the great cultural movements of the past?

A Corrective to the Foregoing Criticism

What has been said so far in this chapter has been preoccupied with the organizational point of view, viz: How does group listening fit into the general classification of adult education? And, from that point of view, it appears that group listening must be content with a modest claim. But anyone turning over again, and from a different point of view, the evidence accumulated in the course of this inquiry, cannot fail to be impressed with the tale of benefits which so many listeners ascribe to their experiences of group listening. While, as a system, group listening may reveal many inadequacies, and while it may fail to improve on existing facilities for adult education, it has unquestionably proved of value in itself to many groups and many individuals. This "consumer's opinion" will be more fully considered later on; but meanwhile it deserves inference and illustration in this preliminary chapter. Consider, for example, the following evidence offered by British witnesses:

1. A group leader writes about the interest his members have been taking in a series called "Children at School," in the course of which some very sharp criticism has been expressed by the broadcaster of certain deficiencies of the British system. He continues: "At the last meeting members were very anxious to know what practical steps might be taken to remedy the defects brought to light in the broadcast—e.g. the large classes in schools, poor equipment, obsolete buildings, and so on."

2. Another leader, writing of the same series, says: "My primary reason for starting this Group was to bring parents into my school. The discussions have tended to settle round conditions in this particular school rather than upon the educational system as a whole, but that doesn't matter at first. I hope that this Group will be the nucleus of a Parent Teacher Association."

There, again, is an admirable achievement of a civic rather than a cultural kind. That group may cease to function as a listening group once the particular series is over, but it may have served to stimulate the establishment of a most important kind of social concern.

3. A third leader sends this analysis of the motives and purposes of his particular Group: "This group is a purely private arrangement and uses the talks as a basis for an intellectual evening in pleasant surroundings. More than half the members are well-informed on the subject matter of the talks but like to hear experts on specific subjects with an opportunity of pooling ideas afterwards. The others, while well-informed generally, are not necessarily concerned with the subject matter in their daily work but are interested in listening to the talk and to the opinions of members of the group during the discussion.

   "To sum up, we use the talks as 'a speaker' to open the evening's discussion and we have no real interest in organizing ourselves as a public group. Nevertheless we all genuinely appreciate the opportunity afforded us of hearing first-rate speakers on subjects of topical interest. From our point of view the talks are excellent, and I firmly believe that their future lies in the private house and not in the public hall. Our group, if it can be labelled so, is the compromise between individual and intermittent listening and the specifically organized public group."

Here, again, is another variety of group listening which, however it is classified, is a contribution to the common good. It embodies a motive on which the adult educator may frown, since it offers no prospect that these people will ultimately become enrolled in some kind of adult education; yet their testimony leaves no doubt that group listening has here developed a valuable and important purpose.

Adult educators are only too liable to judge a cultural activity by the standards and methods of their own systems; and this tendency no doubt accounts for a good deal of the emphasis they put upon the limitations of group listening. For them, more than for any other kind of expert observer, is it salutary to consider the abundant testimony of so many listeners that, for all its technical and organizational disabilities, group listening has become an influence in the lives of many types and conditions of men and women. And if that influence is not so single-minded nor so academic nor so continuous as is the influence of adult education in its sterner forms, yet its value is none the less relevant to any assessment of group listening as an educational device in the fullest sense.

Summing Up

The preliminary considerations touched upon in this chapter may be summarized thus:

1. Group listening in the European democracies has proved an innovation of slow growth. This modest development must be attributed, for the most part, to certain very real disadvantages inherent in the group listening principle.
2. Of these disadvantages the principal one is that group listening can provide only a one-track system of discussion. With this disadvantage there is closely linked the question of group leadership. This substitute leadership of ideas projected by the broadcaster is at best a natural disability of the system: how far it has been or can be mitigated will be considered in later chapters.
3. Group listening has not established itself as a well-defined new colony of adult education. It has depended largely for its existence so far on the support accorded it by the adult educational organizations, yet wherever these are already well developed, group listening has not proved a particularly important collaborator,

4. As an educational method there is some risk of the overvaluation of discussion. To be effective, discussion must be based on possession of a body of knowledge by the participants; and group listening often ends by beating the air or in ignorance.

5. The limited appeal of group listening in Europe contrasts with the volume of "serious" broadcasting there. The interest in educational talks, radio discussions and features is considerable in most European countries; so, too, is the demand for such types of educational broadcasting as courses in foreign languages.

6. The special programmes designed for group listening must not be assessed merely from the formal point of view of the adult educator; they are proved to possess what may be called a diffused value, and in this way are shown to be a social influence of considerable and diverse value.

II ▸ THE GENERAL LAY-OUT IN GREAT BRITAIN

Rudimentary Forms of Communal Listening

GROUP Listening in its simplest unorganized form is the informal family group which falls into discussion about something they hear over the wireless. How much of this exists it is difficult to say, but in a recent local survey of the social effects of broadcasting, conducted in a working-class neighbourhood in Bristol, it is stated that broadcasting has supplied

... a vastly wider range of conversation. This was agreed on all sides. One listener gives a picture of his home. "You get the family sat in the house of a night and there's a talk on the wireless. Some one doesn't agree and pulls it to pieces. Then they all 'as a go and gets outside of it. I've known them argue for hours." A question as to how far discussion of wireless programmes was usual was included in the family questionnaire, as a result of which it was found that over 90 percent of the families in the sample taken habitually discussed what they heard.

Another simple and informal kind of discussion group exists where groups of work-mates or frequenters of the same public house or transport service hear a programme and discuss it either on the spot or later at their leisure. A wartime broadcast by Winston Churchill, for instance, forms a most popular subject of discussion in every train and omnibus on the following morning.

Neither of these types of groups can or should be organized. Their value lies in the spontaneity of their reaction to stimulating material. It is the group which is organized to listen to regular radio programmes given at fixed weekly times that is the subject of this Report. But the two types of group are not entirely independent. The encouragement given to measurable discussion almost entirely contributes to the virility of unmeasurable discussion.

Typical Group Listening

These formal groups are of many different kinds. Perhaps a doctor or a parson or a teacher in a village gets a group of neighbours in to discuss a series of talks. Then again there are groups which meet under the auspices of some society with broad social functions—Women's Institutes, Community Associations, Y.M.C.A.'s, Co-operative Guilds, church societies, political parties, and so on. Or there are the groups organized by bodies whose primary function is educational, e.g., Workers' Educational Association and the National Adult School Union. A further category consists of groups which meet under the inspiration of the librarians of public libraries. Finally, there are the groups run directly by local education authorities incorporated in the public adult educational system and frequently led by men and women who are paid for the work. During the six months from October, 1938, to Easter, 1939, 1,453 groups met in connexion with six series of talks; of this number approximately 400 were organized by local education authorities, and of these over 300 had leaders who were paid.

There is no standard group in Great Britain. In composition, groups vary from four or five members of one age, sex, and social class to fifty or sixty of a variety of ages and classes. A typical rural group visited lately consisted of a farmer, a farm servant, blacksmith, millwright, business man, shopkeeper, schoolteacher, shepherd, housewife, railway plate-layer, and station master. A typical town group was made up of two teachers, pawnbroker, marine engineer, retired railway engineer (both these latter having had service abroad), two miners, three laceworkers, two grocers, and a blind organist.

In outlook, groups vary from those whose members start with a common ideological basis—as a League of Nations Union Group—to groups having no common background except that of locality as, for instance, a group in a local prison. In leadership, they vary from groups where the leadership (usually nominal) goes the round of the members from week to week, to those where the chairman is an experienced—and sometimes a paid—discussion leader. The quality of the group is probably more dependent on the quality of the leadership than on any other single factor, and it is significant that in Great Britain the number of groups with paid leaders is increasing. Groups vary in their response to the broadcast talks. Some listen to the talks first and subsequently discuss them for anything from half an hour to two hours. Some, who find the talk rather late for discussion the same night, discuss the previous week's talk first, then listen to the night's talk and break up. Some do follow-up work of a practical nature—such as visiting local sources of relevant material; others do some reading on the topic of the broadcast. Some groups grow so enthusiastic about the subjects under discussion that they wish the B.B.C. to carry the matter further. For instance, one group, after listening to talks about town and country planning last session suggested that the B.B.C. had a duty to stimulate wider interest in the subject by putting on a series of talks called "The Privileges and Responsibilities of the Individual Citizen in Regard to Local and National Government." Pomposity of this kind on the part of their admirers in the educational world is one of the major difficulties with which the B.B.C. programme planners have to contend. The majority of groups probably do not do any follow-up work as groups, the leader being content to have induced the members to face facts and interpretations of situations which may not have been incorporated in their previous thinking on the subject, some groups only meet for series that interest them for some special reason, while others maintain a sort of "Thursday night" continuity whether the subject of the series be economics or art.[1] In general there is a tendency for the groups to be Left Wing—certainly in political, economic and social subjects. This is natural, since those who are interested in ideas usually tend to the Left rather than the Right, and it is only those who are interested in ideas who will go to the trouble of joining discussion groups.

[1] An effort is made to provide roughly similar types of subject on any given night in the two halves of the winter, but this is not always possible.

The British Set-Up Summarized

With regard to the material for the groups, the following was the schedule during the winter of 1938-39 and represents a fair average of the kind of talk provided for groups by the B.B.C.—until the war broke out.

1938 (October-December)	1939 (January-March)
Men Talking [2]	Children at School
Class: An Enquiry into Social Distinctions	Town and Country
The Mediterranean	The Pacific

For the autumn series a bare statement of the general subject together with dates and times was available about the beginning of July, and fuller details, including synopses of the talks, names of speakers, book lists, etc., about the middle of August. For the spring series similar information was ready about the end of November and the end of December, respectively. This material was distributed free to anybody known to be interested and likely to take active part in arranging groups.

Apart from any expenses incurred by local groups the B.B.C. finances the whole organization; it not only pays the radio speakers and the programme staff, but also the salaries and expenses of seven Education Officers (about whom more will be said later) and a small central administrative staff. Moreover, the work is carried on in conjunction with a fairly expensive form of advisory machinery. In each of the seven regional areas there is an advisory committee of ten to twenty members interested in adult educational development, and there is a Central Committee for Group Listening composed of fourteen members, meeting three times a year. Moreover, a sub-committee of the latter body meets more frequently to discuss with the B.B.C.'s programme staff the detailed form of the talks designed for discussion groups.

That completes a rough sketch of listening groups in Great Britain in the last relatively normal winter before the war. In order to understand how the development reached this point it

[2] The nature of this series is referred to later.

is necessary to go back to the beginning of educational broadcasting in 1923 and to consider a dull but necessary résumé of the growth and development of the administration of group listening in Great Britain.

The British Broadcasting Company was formed in 1922, yet as early as 1923 it set up an educational advisory committee, and in the following year appointed an Education Director. In 1923, the ordinary evening programmes included two talks simultaneously broadcast from all stations, each lasting ten or twelve minutes. There was no regular syllabus, and talks were fitted into the programme as occasion offered. Between 1923 and 1927 the talks-policy of what in 1924 became the British Broadcasting Corporation developed in a variety of informative talks by distinguished men of politics, literature, science, music and so on, while a number of talks on public affairs were arranged in conjunction with Government departments. None of these talks was designed specially to provide discussion, and at this period no group listening was contemplated.

Classification Begins

In 1927, talks other than those which could be classified as definitely educational were placed under the care of a separate department, and special programme officers were assigned to the task of arranging series of twenty-minute talks on "serious" subjects at 7.25 P.M. on five evenings a week, and one half-hour talk at 8.00 P.M. In 1928, a committee of distinguished educationists, set up jointly by the B.B.C. and the British Institute of Adult Education, reviewed this educational work and issued an important report called *New Ventures in Broadcasting*. The committee concluded that broadcasting could supplement the work of existing adult educational movements without any risk of supplanting them. It urged that at least an hour a night should be given to educational talks (interpreting "educational" in a fairly narrow sense), the majority of which should be informative material. Some reference was made to the possibility of discussion groups, but talks designed to provoke discussion were not at first envisaged as being the primary adult educational work of broadcasting. The committee recommended that adequate machinery should be set up by the B.B.C. to plan and supervise adult educational talks and to foster public response to them.

A Central Council for Broadcast Adult Education

This organization was, in fact, set up in 1929 for a period of five years. It included a Central Council for Broadcast Adult Education and four Area Councils, each with its Education Officer. (The Home Counties, Wales and Scotland were not included in this development until 1934-35, when three additional Area Councils and Education Officers were appointed.) During these five years the Central Council planned and supervised substantial programmes consisting of five concurrent series of ten or twelve talks, each broadcast at the fixed hour of 7.30 P.M. each evening from Monday to Friday throughout the year, with the exception of the three summer months.

During this period it became plain that the B.B.C. was not going to be able to allocate as much time to educational broadcasting as the 1928 report had recommended, and this realization led the Council almost imperceptibly to choose the "discussible" rather than the informative subject for the talks. This development was fostered by the removal of the ban on broadcasting about controversial subjects, a ban which had been imposed by the Government in the early days of radio, but which was removed late in 1928 after the Adult Education Committee's report had stressed the importance of including controversial material in radio talks. In the winter of 1933-34 there were in the neighbourhood of a thousand organized groups listening to one or other of the five series.

New Administrative Machinery

In 1934 the machinery for stimulating and organizing groups was again revised. The growing demands for varied programme material and the development of radio entertainment made it necessary for the B.B.C. to have greater freedom in planning the evening programme; the programme-planning functions of the old Central Council were abrogated; the B.B.C. took direct responsibility for all evening broadcasts, but worked with adult education advisory committees both centrally and in the regions to keep it informed on the programme and administrative aspects of the work. It dissolved the large unwieldy all-in Central Council for Broadcast Adult Education, and set up in its place a smaller Central Committee for Group Listening. This Committee could do no more than suggest to the B.B.C. suitable talks for groups; its other function was to stimulate and direct the development of groups. Under the new arrangement, as under the old, the Education Officers had the work of stimulating listening groups in their areas; but they were responsible to the programme departments of the B.B.C., although their work was primarily that of educational administrators. Educational administration is a technical field with which the B.B.C. programme officials were fundamentally unfamiliar; and the arrangement was a bad one in theory and a poor one in practice. None the less, in the early stages it was essential for the B.B.C. to take this administrative responsibility since none of the existing educational organizations was prepared to do the necessary hard work. In fact, although group listening had the official blessing of existing organizations, it met with comparatively little practical sympathy. Adult educationists contented themselves (at best) with advice and good wishes. At worst they criticized group listening in twenty different ways—especially by expressing doubts whether the B.B.C. was maintaining intellectual standards.[3]

However, between 1935 and 1937 it became clear that the development of groups would never be soundly based unless the administrative aspects were supervised by specialists with the necessary training and knowledge. Since school broadcasting had already secured the services of a distinguished Director of Education to look after the administrative side of school broadcasting, it was decided early in 1937 to put the administration of group listening under the same supervision. As Secretary of the Central Council for Group Listening he was assisted by a group of educationists, who continued to act as an advisory body on programmes for the B.B.C. and had, indeed, a programme sub-com-

[3] A symptom of the difficulty of uniting broadcasting and adult educational interests was the appointment as B.B.C. Education Officers of men with little practical experience of adult education outside universities. Of the first four appointed only one was an adult educationist by profession.

mittee which continued to work in close contact with the responsible programme officials.

The B.B.C. officers were now free to devote their whole attention to programmes, while experienced educationists (i.e., the Central Committee for Group Listening) were able to go ahead with the work of organizing groups and training leaders. The B.B.C. continued to provide the finance for the whole organization, which was to continue until 1940, by which time it was hoped that the educational organizations of the country could take a greater share of the cost of organizing and supervising the groups. The war has, however, played havoc with this foreshadowed development, and owing to the curtailment of programme alternatives for defence reasons the whole system of talks-broadcasting has had to be modified almost out of recognition.

A Summary of This Administrative Development

The evolution of the group listening movement in Great Britain has thus seen several changes of constitution and the B.B.C. personnel has changed frequently. In an organization growing as fast as the B.B.C. has done in the last fifteen years, it was impossible to hope that the same officers would remain in charge of this particular aspect of the work, and the continuity of the B.B.C. policy has suffered accordingly. But on the advisory side the personnel has remained very constant. The Central Committee for Group Listening is a body which remains unchanged and continues to represent the major interests of British adult education. It consists of one representative each of the British Institute of Adult Education, the Workers' Educational Association, the Association of Education Committees, and the Universities' Extra-Mural Consultative Committee; one representative of each of the seven Area Councils covering England, Scotland and Wales; and three co-opted members representing particularly the interests of women, rural listeners, and tutors engaged in adult education. The Area Councils, which exist to stimulate group listening in their regions, are drawn from the ranks of educationists, librarians, teachers, clergy, and group listeners. Their executive officers are the area Education Officers of the

B.B.C., who are also charged with the responsibility of encouraging schools to take the school radio service in their areas.

An important feature of the present organization is that it shares all save one of its principal executive officers with the Central Council for School Broadcasting. It has been found that this arrangement has much facilitated the approach to local education authorities and national organizations who are concerned with the whole field of educational broadcasting. It reflects also the growing recognition that the two sides, junior and adult, are closely and naturally linked.

There is, however, an important difference in the functions of the two bodies. Unlike the School Broadcasting Council, which is responsible not only for organizing the "listening end" but also for the planning and supervising of the special schools programme, the Group Listening Committee has no executive capacity in the programme sphere. Three concurrent series of evening talks are provided by the B.B.C. with a special view to the needs of listening groups, but these are, of course, intended for the individual listener as well. They come, therefore, within the framework of the talks-programme as a whole, in regard to which the B.B.C. is advised by its Talks Advisory Committee. The Group Listening Committee has, however, set up, under its constitution, a Programme Sub-Committee, qualified to advise on the special needs of listening groups, which has two representatives on the Talks Advisory Committee and is, moreover, informally consulted by the B.B.C. as regards subjects and speakers likely to appeal to listening groups. The design of the talks for discussion groups is thus also in part, like the Schools Programme, in the hands of those for whom they are more especially intended.

The eleven-years' development from 1928-29 to 1938-39 reveals two interesting and significant features. The first feature which has developed in this eleven-year period is the fact that talks for groups have increasingly become provocative rather than informative material. The second is the firm differentiation of function between the B.B.C. as the provider of talks for groups and the Central Committee as the organizer of the consumption end. This firm separation of function has the merit that each of the two collaborating interests knows where it is. Some members of the Central Committee believe their functions are now merely nominal, since the B.B.C. always has the last word about what talks shall be given or not given. On the other hand, although the B.B.C. makes this final decision, it does receive suggestions for talks from the Central Committee, and very often adopts them. This demarcation of roles goes so much to the root of group listening that it will be more fully explored in the next chapter.

Postscript on the Situation Since War Began

With the outbreak of war the whole broadcasting service had to be rapidly adjusted to meet a new situation. National and regional programmes as planned were at once discontinued, and were replaced by a single Home Service programme. *Pari passu* there was a great increase in the output of news and directional talks in foreign languages. There was an equal disturbance in staff arrangements. Many departments were moved into the provinces; many individuals were transferred to new departments. The talks which had been specially arranged with listening groups in mind, in common with much else in the programme, were excluded for the time being. During the first month or two, a considerable proportion of the peak period in the evening between 7 and 9 P.M. was occupied with news bulletins, talks by Cabinet Ministers and notices of national and regional importance. The rest of this period was given over largely to light entertainment, partly because the most popular forms of programme must predominate in a limited period, and partly for the purpose of contrast to news and notices and for general flexibility. Talks on general subjects, therefore, at these times suffered an eclipse, and it was not until the need for notices became less urgent and they could be cleared out of the way from 7 to 8 P.M., that room could be found for a greater number of serious items such as music, plays, and talks.

Correspondence received by the secretary of the Central Committee for Group Listening from all kinds of educational organizations as well as from the more serious individual listeners was very impressive, and helped to confirm the views of the programme planners that there is a large evening audience for serious talks. The B.B.C. accordingly decided in early November, subject always to circumstances permitting, to broadcast two additional serious evening talks each week at times that would be suitable for the type of listener who is disposed to join a group. It should be stressed, however, that the B.B.C. adheres to the view that as long as there is only a single programme, particular interests cannot receive separate consideration (especially in the evening). Every programme must justify its position on the grounds that it is making a wide appeal. Talks for listening groups, therefore, do not occupy in the present programme the separate place they had in pre-war days. It is true that "The Artist in the Witness Box" proceeded as it had been planned before the war began, but both the series "Europe in Travail" recently broadcast by Middleton Murry, and the series "Men of the Hour" have had the general listener in view. The Corporation, however, deliberately placed these talks in a part of the evening (i.e., between 7 and 9 o'clock) where the large audience comprising those interested in adult education in some form, whether listening in formal or in fireside groups, or as individuals in their homes, could take them.

Information about these talks was circulated as widely as possible through Education Officers, local education authorities, adult educational organizations and the educational press. At the same time, there were picked out from the general programme other items such as plays, music, school broadcasts and general talks which seemed likely to interest the group listening public.

During the first session of 1940, 285 groups were known to be still listening—a figure which is only slightly less than half of the usual average figure per session. These known figures are certainly an understatement. Education Officers are limited in the amount of travelling they can undertake, and in visiting groups they are particularly affected by the blackout, so that exact figures are impossible to ascertain. Then, again, the organization was handicapped by the fact that the B.B.C. did not restore "serious" talks until early December. Large numbers of groups which had arranged to follow the series planned to start in October had to switch over to some other form of study, or found themselves busy with A.R.P. or war service. During the first two months of the war, programmes were not planned far ahead. Even yet the long-distance planning of pre-war days has not returned. The result has been delay in getting information to the Education Officers. This in its turn added to the difficulties of local groups in getting things going at short notice. There must be very many informal groups meeting round the fireside about which very little is known.

III ▸ THE FUNCTION OF ADULT EDU-CATIONISTS AND THE FUNCTION OF BROADCASTING

ASSUMING for the moment that the policy of broadcasting provocative talks with a view to discussion by the audience is justified, there are two main conditions of success:

(a) people must find it desirable and possible to engage in discussion; and

(b) there must be the right material to discuss.

The Special Rôle of the Educational Bodies

The fulfillment of the first of these is the job of educationists, and is linked up with the problem of providing group leadership. Any effective discussion involves its members in thinking, and systematic thinking is not a natural activity of the Englishman, even if it is of the Scot or the Welshman. Thinking involves a hard moral decision, and as a rule can only be provoked by the personal contact of mind with mind. Wireless does not readily encourage that sort of discipline; the listener who finds a talk too "difficult" is prompted to switch off and seek something which makes a lesser demand upon his attention. The cultivation of the "listening discipline" is a process which adult educationists alone accomplish among the listening public, and they can accomplish it only through the medium of group listening. Already the adult educationists have demonstrated that this is their *métier* in educational broadcasting.

Thus, in the early days of group listening when the B.B.C. was still responsible for the listening end as well as the providing end, it held expensive summer schools at Oxford with a view to training group leaders. The results were meagre; but when at a later stage this responsibility for training group leaders was accepted by educational bodies, a very much larger number of schools' and leaders' courses was arranged for shorter periods in a wide variety of locations; in the winter of 1937-38 no fewer than seven hundred potential leaders were gathered together in about forty short courses for training in the methods of group leadership, and in the first winter in which the educationists took administrative responsibility for encouraging group listening, the numbers jumped by approximately fifty percent. The desire to discuss must continue to be the particular care of the educationists concerned with this work. The material for discussion, on the other hand, must be the responsibility of the broadcasting authorities. There are obvious reasons why.

Control of Programme Planning

In order to secure the maximum number of groups it is essential that the talks for groups should be put into the programme at popular times. Yet it is at these times that there is also the largest number of listeners who are not interested in talks for discussion. This means that the radio authorities must take full responsibility for what they are doing, i.e., satisfying a minority interest at a time of potential majority listening. The B.B.C. cannot, therefore, delegate its responsibility for programme planning to an outside group of people. It must maintain complete control of the matter and form of broadcast talks, attempting to satisfy at the same time those who require provocative talks for purposes of discussion and that much larger body of listeners who enjoy talks of good calibre, but who do not want to carry their listening to the stage of turning out of their house to discuss what they have heard with a lot of other people.

Where Adult Education Went Wrong

As the function of the two responsible partners has become more clearly differentiated, it has become clear that there is a good deal of fundamental misunderstanding between the two. The organized educationists, who are the only people from whom the B.B.C. can draw such advisory committees as the Central

Committee for Group Listening, are brought up and trained in the academic tradition; as tutors and organizers in the W.E.A. and extra-mural departments of universities they tend to think of "serious" talks in terms of modified university work. Even when they bring themselves to contemplate the problems of the beginner in adult education, they are still thinking of how to bring beginners to university and near-university standards. British adult education is still damnably myopic about really elementary educational needs. Moreover, for a variety of reasons, the adult education organizations are sensitive about their "running rights" in this field.

It is difficult to analyse the calibre of much of the adult educational work conducted on traditional lines, but in many instances it is certainly true that it is not as good as its organizers would in their heart of hearts like it to be; and this makes them touchy about the work of those whose educational purposes are different and possibly more popular. Organized adult educationists have always been fundamentally suspicious of the B.B.C.'s honesty and ability in fulfilling its educational responsibilities. This suspicion has not necessarily been deliberate. For the most part the educationists have had experience among people who in one way or another have already been aroused to a relatively abiding interest in the things of the mind. At any given time there are perhaps a million people in the country who have pursued, are pursuing and will pursue some further educational course of a nonvocational type; the audience for a good radio talk of the kind that is put into the discussion group periods is in the neighbourhood of five to seven millions.[1] The educationists have experience, and think primarily in terms, of the élite of a quarter-million. While one of the weaknesses of broadcasting in the past has been the B.B.C.'s unwillingness to recognize the importance of what may be termed "directional broadcasting," it is none the less true that the radio programme officials are, and ought to be, more concerned with those lower down the scale of intellectual interest than with those at the top, for the latter can pursue their studies and interests through the medium of classes and reading.

[1] The figure is supplied by the Listener Research Section of the B.B.C.

A Celebrated Series

The educationists have always pressed the B.B.C. to arrange its programme in the traditional pattern; and the best example of this may be seen in "The Changing World" series in the winter of 1931-32, when the following schedule of evening talks was provided:

Sundays
5.00–5.30

The Modern Dilemma (24 talks)

Christopher Dawson, Prof. John Macmurray, T. S. Eliot, and others

Mondays
7.30–8.00

Industry and Trade (24 talks)

How Wealth Has Increased	Prof. Arnold Plant
Why Does Poverty Continue?	D. H. Robertson
How Has Private Enterprise Adapted Itself?	Prof. Henry Clay
How Has the State Met the Change?	Prof. Arnold Plant

Tuesdays
8.30–9.00

Literature and Art (24 talks)

The New Spirit in Literature	Harold Nicolson
The Drama	Sir Barry Jackson
The Press	Kingsley Martin
Modern Art	J. E. Barton

Wednesdays
7.30–8.00

Science (24 talks)

What Is Science?	Prof. H. Levy
What Is Man?	Prof. J. Huxley
	Dr. John Baker
Science and Civilization	A Symposium
Science in the Making (II): Changes in Family Life	Sir William Beveridge

Thursdays
7.30–8.00

The Modern State (24 talks)

Can Democracy Survive?	Leonard Woolf
	Lord Eustace Percy
Diseases of Organized Society	Mrs. Sidney Webb
Has Parliamentary Government Failed?	Prof. W. G. S. Adams
The Problem of World Government	Sir Arthur Salter

Education and Leisure (24 talks)

Fridays 7.30–8.00	Learning to Live	Prof. John Macmurray Prof. J. Dover Wilson Sir Percy Nunn
	Modern Life and Modern Leisure	Prof. C. Delisle Burns

On paper this was a magnificent plan as an introduction to what was going on in the contemporary intellectual world. In educational circles the series was a great success. A very large number of the contributions were subsequently printed or incorporated in permanent book form and have played an appreciable part in moulding the intellectual life of the decade. The series was valuable to the B.B.C. because it gave the educational and intellectual world confidence that the B.B.C. was prepared to take responsibility for far-reaching analyses of the contemporary world, and without the support of the intellectual world the B.B.C. would have found it a much more difficult, if not impossible, task to stand up in following years to the vested interests which consciously or unconsciously were anxious to curtail the power of broadcasting to stimulate free thought. In the immediate purpose, too, of stimulating group listening over a single season this series was good. But from the point of view of radio-talks policy in the long run the series probably did much harm. The person to whom the broadcast brings something new and exciting is not as a rule the person with a trained mind which can feed upon printed material and upon conversation with equally able friends. The talk should rather be directed to those who without it would be starved of a vital form of mental stimulus. Radio should try to bring within range of disinterested thought and analysis those millions of ordinary listeners who are interested in the world about them, but who for one reason or another lack the background which will enable them to satisfy their curiosity and concern through books, good periodicals, and informed conversation. For such people this famous "Changing World" series was meat far too strong. The material was too difficult and often too abstract, the thought too condensed, and the whole intellectual scope too vast to be grasped by their imagination. Moreover, so massive a series as this crowded the programme so much that anyone tuning in to a talk was much more likely to find himself listening to some aspect of a broad analysis of current intellectual life than he was to find himself listening to the homely and personal talk, grave or gay, telling of how the speaker had coped with his personal environment.

This celebrated but ill-judged series won for radio talks a reputation for solemnity and abstraction which was probably not deserved even then, and has certainly not been deserved since. But the prejudice of many casual listeners toward talks probably dates from this time when policy was dominated by the magnificent tradition of British adult education.

The Growth of a New Talks Tradition

Both the B.B.C. and the educationists have always been proud of this series, as indeed they were entitled to be if the premises on which it was based are valid. They have even asserted that the high quality of such series as "The Changing World" served as a Maginot Line, i.e., as an assurance that there will always be *some* serious talks in broadcast programmes. The educationists have tended to cling to the tradition which produced "The Changing World" schedule. But the B.B.C., on the other hand, has perceived its weakness, and for the past nine years it has been steadily experimenting in new techniques of presentation, usually in face of the criticism of the majority of educationists, though with the support of the more adventurous of their number. Some of the modifications introduced by the B.B.C. were the shortening of talks from half an hour to twenty minutes, the introduction of many more "lay figures" as witnesses to produce evidence from everyday life, the introduction of the man in the street to act as interrogators of experts and, finally, the adoption of an extremely loose and conversational form of broadcasting.

"Men Talking," for instance, was a programme in which three or four people began talking about a subject in a studio before the microphone was switched on. When the programme was due to begin the microphone was hooked up and the conversation continued until the end of the allotted time, when it was slowly faded out. One of the speakers was sometimes an expert on the subject under discussion, but the whole performance depended upon the completely spontaneous nature of the conversation. It was the sort of thing which might be overheard any day in a railway carriage where three or four people with reasonably ready minds and tongues found themselves casually collected. The topics on which they were invited to speak were usually relatively important matters coming within the personal experience of vast numbers of listeners. The first series dealt with questions of family life; the second with various common aspects of social life, e.g., tipping, sense of humour, professionalism in sport, and so on. This series was anathema to established educational interests, but won a larger listening public than almost any series of talks broadcast up to that time. Technically, "Men Talking" was not entirely satisfactory, and a good deal of experiment would have been necessary before it could have been regarded as a wholly satisfactory method of broadcasting from a listening point of view. But at any rate its success as a method of stimulating critical interest entitled it to much more experiment, which would doubtless have been forthcoming but for the war.

The 1938-39 schedule [1] shows how far the subjects had developed from the beginning of the decade. The Report of 1928 (*New Ventures in Broadcasting*) mentioned international affairs as a useful field of study, and the series on the Mediterranean and the Pacific were, of course, highly topical in the troubled year before the outbreak of the present war. Yet these two series are the only two in the list which would have fitted into an ordinary educational syllabus. "Men Talking" is a technique which can only be applied to radio. The series on "Class: an Enquiry into Social Distinctions" could, again, only be done by broadcasting, where the treatment and the illustrations made it possible to present the problems with immediacy and topicality, yet without the offensiveness which would have been almost inescapable had similar talks been delivered face to face with an audience of very mixed social origin. "Children at School" was a series examining current educational practice, which drew upon a variety of parents' and teachers' firsthand evidence that probably only radio could collect and present. "Town and Country," an examination into some of the sociological problems of the distribution of

[1] See p. 169.

population in Great Britain today, was also drawn on a wider and yet more personal scale than any class would be likely to achieve for itself: a housewife previously living in a slum and now living in municipal flats could talk about the old life and the new with a freedom and unself-consciousness which would have been quite impossible by any other means than radio. The subject as it was treated by radio could never have figured in any adult educational syllabus in Great Britain.

Educational Talks, Informative or Provocative?

This comparison between group listening programmes and the contents of an adult education syllabus must not be taken to mean that the adult educationists were uniformly hostile to anything the B.B.C. attempted to do. As a rule each new step taken by the B.B.C. was regarded in advance with suspicion as a possible lowering of standard, but was applauded in retrospect as it was seen that the method had succeeded in widening the circle of interested and critical listeners. This picture perhaps paints the B.B.C. rather too white as the progressive element in the partnership. It is true that from the start the B.B.C. has recognized its educational responsibility, has spent much money on it, and has usually been responsible for devising new and successful techniques. But the B.B.C. has not always convinced its adult education partners that it was participating with enthusiasm in the whole plan. From a crude programme-planning point of view there are three objections to the less academic type of material in recent use:

(1) It is far more difficult to arrange a series of ten provocative talks on the desirability or otherwise of class distinctions or on the nature of progress than it is to arrange a series of informative talks on British history, or on the growth of scientific discovery. Informative material is always easier to produce than a balanced controversy involving the fair presentation of different points of view. Moreover, it is probably true that the temper of the world in the 'thirties as a whole was much more disposed toward the lapping up of facts than it was toward the consideration of controversial issues, and the B.B.C. "Talks for Groups" schedule could certainly have attracted a larger public had the same amount

of preparation gone into presentation of informative talks as went into provocative talks.

(2) Obligation to the organizers of listening groups imposes absolute rigidity of programme planning so far as talks for groups are concerned. There can be no juggling with days and time to accommodate more popular items, a practice which would be possible if talks were informative only and were intended for individuals listening by themselves at home.

(3) No educational administrator will embark on a policy unless he can be reasonably sure that he will be able to control the conditions of its development for some years to come. Five years for the development of a new educational project like group listening seems a comparatively short span to an educationist. Yet to a radio corporation, particularly at an early stage of its growth, five years is likely to see a whole string of major changes of technique and policy: indeed, it is probably true that from 1928 to 1936 technical developments in the presentation of entertainment programmes were far more rapid than in the presentation of talks programmes. The consequence was that the comparative appeal of talks declined, and so, therefore, did the number of occasions on which they could justifiably appear in the daily schedule.

For all these reasons the B.B.C. has shown signs from time to time of departing from its own high professions, and on more than one occasion there has been good reason for supposing that it would be glad to be able to pursue a policy based on its own convenience and untrammelled by the advice and comments of its educational advisers. If the educational advisers have done little to encourage the technique of presentation to groups, they have been invaluable in bringing pressure to bear on the Corporation to persist in a difficult but important form of social responsibility; and if the B.B.C. has not always appreciated the efforts of its educational advisers to keep it on the straight path of stimulating democracy, it has shown considerable fertility in developing techniques which have introduced to the British public as a whole wider aspects of contemporary life than would ever have been open to them in the past.

IV ᵥ ORGANIZATION AND METHOD
OF LISTENING GROUPS

THE responsibility for selecting the subjects and the speakers is, in the last resort, the B.B.C.'s; although as shown in a previous chapter the B.B.C. is guided in its decision by the views of the group of adult educationists who sit on the Central Committee for Group Listening. The B.B.C. issues a preliminary pamphlet before the opening of each of the two annual sessions, viz: the spring session (January to March) and the autumn session (October to December).

Who Creates the Groups?

On the basis of this announcement the business of organizing groups begins. The primary responsibility for this lies with the seven area councils which cover England, Wales, and Scotland: councils which include the whole range of representatives of adult education bodies, and each of which possesses a B.B.C. Education Officer who divides his time between work for this area council and for a similar one for school broadcasting.[1] A considerable burden of organizing falls upon these Education Officers. They address meetings, organize week-end schools and conferences, visit possible supporters, advise on how to start and maintain groups, supply information about forthcoming talks and answer inquiries of all sorts. The Education Officers have very large areas to cultivate; the Home Counties' Area, so called, stretches from Norfolk to the Isle of Wight and the North-Eastern Area from Northumberland to the Wash. Moreover, the officers devote only half their time to this work, the other half being directed to the schools side. Nevertheless, they manage personally

[1] See Chapter VI.

to visit a large number of groups and listen to the talks in their company; their reports on how the talks are received are of the greatest value, both to the group listening organization and to the B.B.C. On their power of inspiring enthusiasm and smoothing out difficulties the formation of groups up and down the country largely depends.

But the organization of groups does not by any means rest solely, or even primarily, upon these Education Officers. Each Area Council, as we have seen, is composed of local officials of the various adult education organizations, who therefore have from the outset an interest in promoting listening groups within their own particular organization. The whole range of these bodies, therefore, plays some part in establishing groups within its area— the Local Education Authority, the Workers' Educational Association, the Y.M.C.A., Y.W.C.A., Women's Institutes, Rural Community Councils, Townswomen's Guilds, Adult Schools, and so on.

The most vigorous of these promoting agencies nowadays is the Local Education Authority which in many instances has come to accept the listening group as an integral part of its provision of public education. The latest available figures of group listening are these:

	Number of Listening Groups
Autumn Series, 1938	
3 concurrent series of 12 talks each, running from October to December, plus one additional series of discussions for the "Under Twenty Club" groups	714
Spring Series, 1939	
A similar schedule running from January to March	656
Summer Series, 1939	
A supplementary service, consisting of a single series of 12 talks from April to June	83
Total	1,453 [2]

[2] The precise number of individuals in this total cannot be ascertained, but can be estimated as round about fifteen thousand. Many of the members of the autumn groups were also members of the spring groups; and the net number of

Of this total number of listening groups nearly one-third are organized by the Local Education Authority—a proportion which increases year by year, largely owing to the recruiting activities of the B.B.C.'s Education Officers who, through their work in school broadcasting, are already in touch with local education authorities.

In this work, as in other educational developments, the local education authorities have revealed a notable sense of realism. They attempt to diminish the dangers of indifferent "amateur" group leadership by instituting as paid leaders teachers employed in their day schools. Eighty percent of Local Education Authority Listening Groups have this kind of paid leader. The rate of remuneration ranges from 5s an evening to 12s 6d an evening, but it seems to be just enough to turn the scales and to attract teachers to the work. The South Wales Council of Social Service and the Gloucestershire Rural Community Council are the only bodies to emulate this realistic policy of the local education authorities.

There is, in all cases, a considerable carry-over of members from the autumn series of talks to the spring series, and there are scores of groups which have run continuously for three or four years. Although the audience for group listening is largely a casual one, it is developing a hard core of "regulars." As a rule, however, these listening groups regulars are also regulars of the particular adult education body with which they are also associated. There has not yet developed anything that can be called a continuous and separate group listening movement; and, apart from the alarm which such a development might cause among adult education organizations, the B.B.C. has not the least desire to foster such an independent movement. Group listening has come to be regarded, by all its partners, as primarily an auxiliary and supplementary activity of the established bodies of adult education.

Listening groups meet in a wide variety of premises and listening conditions; this matter is examined in another context in Chapter VII, where it is considered as one of the four crucial factors in promoting group listening.

_____

individuals who shared in group listening for the whole year can be reckoned at about ten thousand.

Pamphlets

In the matter of pamphlets and other publications for the special use of groups the B.B.C. is liberal and generous. Preliminary details about forthcoming talks are circulated to interested organizations and individuals some three or four months beforehand. Later, a full descriptive pamphlet is issued (free), containing names of speakers, synopses of the talks, lists of books for supplementary reading and, in certain cases, suggested questions for discussion. From time to time one of the series of talks is selected for even fuller treatment in the form of a special printed pamphlet. One outstanding example was the pamphlet "Design in Everyday Things," issued in October, 1937. Even more impressive was the beautiful booklet issued by the B.B.C., after the war had broken out, as a guide to the series on "The Artist in the Witness Box." It was a collection of fifty illustrations, mostly in colour, of paintings and sculpture; and the issue of such a sumptuous publication at a moment when Great Britain was again embarking on a colossal war, lends reasonable colour to the view that this country is concerned with preserving the cultural as well as the material benefits of civilization from the German tyrant and vandal.

How Groups Work

In the special pamphlet issued by the B.B.C. for the guidance of group leaders there is a wise avoidance of any hard-and-fast directions for the conduct of listening groups. Answers to a questionnaire submitted to group leaders during this investigation reveal an impressive variety of method. A verbatim record of their testimony would be repetitive on many points; but the following abstract of points from these questionnaires gives a good idea of the average procedure:

Preliminaries

Most groups make a practice of assembling a quarter of an hour or so before the talk is due to begin. Generally this period is occupied by an exchange of reflections upon the preceding talk in the series. Sometimes a member of the group reads a brief synopsis of the previous talk—a contribution which is materially assisted by the fact that most talks are printed (after they have been given) in the B.B.C.'s weekly journal *The Listener*. If the series is one on such a topic as international affairs, atlases or wall maps are made ready; in other cases the illustrative material provided by the B.B.C. (i.e., pamphlets) are distributed to the group. This preliminary period is also, but more rarely, used by the leader to sketch the broad outline of "tonight's topic." Many groups, however, appear to object to this, reporting that it is necessarily a tentative anticipation of what the Speaker may say, and is therefore liable to confuse the issue. Some leaders, without attempting such an anticipation, give a brief introduction of the general issues which are likely to be raised. But for whatever purpose this preliminary period is used, it is always brief. Some rare instances occur in which the group listening activity appears to be merely an item in a two-hour session. The proceedings begin with a talk by a speaker on a topic parallel to the one which is presently to be broadcast. This talk is followed by a brief discussion; then the wireless talk is switched on; after which there is a comprehensive discussion covering both the speaker's address and the broadcast one. Few groups appear to favour this mingling of methods—and for sound reasons. If the wireless talk were merely a solo interpolation of special evidence it might fit well enough into such a pattern. But the talks are often a two-part or three-part debate; so that the multiple method practised by this small minority of groups becomes confusing to the participants. General experience suggests that the best preliminary to the evening's work is the brief summary of last week's talk, reinforced by a few explanatory headlines (so to speak) about the talk due to begin in a few minutes' time.

Receiving the Talk

The second stage is the reception of the broadcast which now-adays never lasts longer, in Great Britain, than twenty minutes; and it is the usual practice for the leader to make absolutely sure that the radio is properly tuned in before the talk begins. This adjustment is of the utmost importance.[3] There is a wide variety of practice in the manner of hearing a talk. Some leaders believe

[3] See also Chapter VI.

in note-taking by the members, and encourage the habit by asking the members to sit at tables provided with paper and pencil. Many of these disciplinarians point out that they adopt this practice on wider educational grounds and not merely to ensure accurate impressions of the particular talk in progress. Other leaders assert that note-taking often interferes with really receptive listening—a contention which may very well be true of listeners who are not accustomed to putting things on paper.

Leading-in to the Discussion

Most groups make a pause in the proceedings as soon as the talk is over. Quite often the incorrigible British habit of tea-drinking is given a chance to assert itself at this point—a practice, in fact, which is a distinct encouragement to informality. As a rule this break is followed by the preliminaries of discussion— the leader clears up incidental difficulties such as an unusual word used by the broadcaster, or something of that kind.

The Group Discussion

The fourth phase is the one in which occurs the widest possible variety. Its nature depends not only on the quality of the leader but on the knowledge, the education, and the interest of the members. Examples are far from common in which this period is a doleful and unproductive affair, a pitiful sequence of bare questions and barren answers. This kind of group, mercifully, is the one which soon peters out of existence. In better cases, the leader selects from the broadcast a few major points which seem to him crucial and provocative; and these, at the outset, he states rather than examines. If his selection is sound and if his group is good, the leader's first task is effectively accomplished, and for the next half hour his business is merely that of chairman. An alternative device is for the leader to ask the group to state what they regard as the key issues of the talk, and to select from these suggestions one or two major topics for the subsequent discussion. Some of the keener groups provide a stimulus to their deliberations by inviting a special guest with particular knowledge of the topic—a schoolmaster if the talk is about education, a farmer if the subject is agriculture. The occasional and discreet introduction of a special witness of this kind is found to be a popular device, and is favoured by at least a third of the British groups.

One group, listening to a series of 12 talks by W. E. Williams on "Children at School" invited guest-speakers on five evenings:

(a) two senior school-girls who took part in the discussion on home work
(b) the chief inspector of schools
(c) a games mistress from a local school
(d) the Chairman of the Local Education Committee.

Of another group listening to the same series, a visitor reports:

Last night's talk was "Taking Their Hurdles" [a talk on examinations], and Mr. Davis [the schoolmaster group leader] had an exhibition of examination papers of different types, intelligence tests, and even some very interesting piece of apparatus which had recently been devised to test mechanical aptitude. He intends to give the group an intelligence test some time, and they are looking forward to it.

One thing that struck me about the discussion—and made me regret all the more that it was so one-sided—was that it was something quite different from the kind of discussion one gets on foreign affairs, etc. It was a Council of teacher and parents discussing John, Mary, Tony, and their individual problems in relation to the subject of the talks, and this made it seem unusually vital and real.

Another group leader writes:

During the Mediterranean series we had great help from a young man home on leave from the Nigerian Civil Service; on the night of the talk on Australia and the Islands—a university graduate fresh from New Zealand (female) who was passing through Keswick saw our poster and was most interesting as she has specialised in anthropology. We've had commercial travellers who had just returned from Italy and Palestine.

Some broadcasters attempt to assist group leaders by formulating three or four definite points for discussion at the end of the talk. Leaders differ as to the value of this practice. Some of them assert that such cut-and-dried discussion items give an element of artificiality to the proceedings, and that the topics for debate should develop organically from the interest and opinion of each separate group. On the other hand some leaders—possibly the less efficient ones—seem to favour this provision of points-for-discussion.

In nothing is the group leader's skill more revealed than in his judgment of how and when to conclude the proceedings. The discussion seldom lasts from more than three-quarters of an hour or an hour; when it exceeds that span it usually does so because the leader is inexpert and inadequate. Several leaders report that they attempt to reach a "decision" on the discussion, often by submitting some sort of resolution. This method, they claim, gives the group the feeling that it is "getting somewhere" and not just debating. There evidently are many people who share this feeling, and for an elementary group there is a good deal to be said for the method. But it is rarely practised in Great Britain, where the usual procedure is for a brief final summing-up by the leader of the whole evening's discussion.

Further Reading

Some groups take seriously the "Suggestions for Reading" included in the pamphlets published by the B.B.C. in connexion with the larger series. Groups which meet in institutions are often equipped with a library containing the necessary books for reference and further study; and where this is the case the leader usually makes a point of guiding the group in the matter of reading. But it must be admitted that only a minority of groups take their reading seriously. A simple investigation of 123 British groups showed that only 16 of them read systematically on the subject of the talks. No reading at all was reported from 33 groups; while the remaining 74 report desultory reading by some of their members. One enterprising group reported having its own bookshelf of "sixpennies" from Penguin and Pelican series—which seem to have been in keen demand. Another group reported that one of its members took it on himself to provide the group each week with a select list of suggestions for reading from relevant periodicals.

County libraries and public libraries appear to be willing to supply books for discussion groups on demand. Some public libraries keep a special shelf of books recommended for current wireless talks and discussions.

Other aids to study—such as the cinema—are almost entirely beyond the reach of discussion groups. Only 3 of the 123 specially investigated had ever arranged a visit to an appropriate film. The main deduction to be made from this is that the cinema is, lamentably enough, very rarely used in adult education in Great Britain—where even its use in schools is far below the average use in any of the larger countries of Europe. Some modest co-operation has occurred with museums and art galleries, some of which have staged exhibitions to illustrate art talks. But there has been a totally inadequate initiative in this matter.

"Social Purpose" in Group Listening

Some zealots of adult education in Britain are inclined to rate the value of any specific type of adult education by its power to quicken social activities among its members. This evaluation is a dubious one, and in its extremer form does not appear to be taken too seriously by the groups. In its more reasonable mani-festations, however, it is frequently discernible among groups. Thus a rural group had arranged visits of its members to several new types of educational institutions; another group had formed a Parents' Council to carry further certain recommendations made in a series of talks on "Children at School." A third, after listening to a series on "Unemployment" had organized a social centre for the wives of unemployed men. Many groups, further-more, report that members had been roused to look more closely into matters of local government and had exerted the pressure of their opinion in matters of local civic interest. On this matter of purpose one group leader very pertinently writes:

Corporately I don't think a listening group can *do* anything with-out the risk of becoming another "front," "wing" or "party." It can only *do* things through the individual so that in the end 70% of the value of the group lies in listening to the expert and discussion; 30% of the group's value is probably lying dormant in many cases. This potential value will only become real as group leaders are capable of making individual members aware of their responsibility and of openings for the exercising of individual effort.

Records of the Discussion

One feature which has been considerably developed in the British adult education movement is the practice of getting students to do a certain amount of "written work" on their subject of study. The effort is rightly regarded as a method of encouraging systematic study. In so elementary an educational mode as group listening there is less emphasis on this academic discipline. Yet some groups have developed the practice in a rudimentary way. Many of them encourage note-taking, but a few are more ambitious. Some of these appoint a panel of their members to draw up a report of the whole series and this report is issued, in stencil form, at the end of the series. Some others keep a logbook, in which, week by week, one member of the group writes up a brief record of the last discussion. This record is read at the beginning of the next meeting and serves as a useful "refresher" for the discussions.

The Consumer's View Again

The rank-and-file opinion about groups is on the whole favourable; but the following selection of comments and experiences includes a just proportion of criticism as well. A number of witnesses were good enough to put on paper their experiences of group listening. These exhibits will help to give a picture of groups at work:—

A School Master's Testimony

It was after attending a meeting at Norwich three years ago that I realized the excellent possibilities of Group Listening and Discussion, and as I was Principal and Organizer of an Evening Institute with nearly 100 young men and women on its roll, I felt that a wireless discussion group with its informality and friendly atmosphere should prove most attractive to our members. I was not disappointed, and not only did members of the Evening Institute join the group, but several persons who had previously shown no interest in our classes were attracted to the listening group, and as a result joined other classes under the Evening Institute.

Our group, which has met 48 times under the auspices of the Blundeston Evening Institute and East Suffolk County Education Committee, is in its third year as an organization. As a preliminary

I wrote personal letters to a few people and invited them to a meeting to talk things over. I also exhibited a public notice inviting all interested. About half-a-dozen people turned up, and these proved ambassadors, for at the next meeting the number was doubled and we formed our group. Those present included the President and Secretary of the local Women's Institute, the Secretary and Chairman of a Young Farmers' Club, the Chairman and Secretary of our Evening Institute Amateur Dramatic Society, and representatives of our Debating Society. There were three housewives, three school-teachers, two gardeners, a clerk, a typist, an agricultural worker, and our ages ranged from 18 to 70. We elected a librarian, a secretary and a leader, the latter position falling to myself. Although I am not prepared to say that the school is the ideal place for a group to meet, the fact that the group was unanimous in its wish to make the school our meeting-place speaks for itself, and it also demonstrated that the object of the organizers of adult education—that of making our school a centre of social and cultural education for adults as well as children— was being fulfilled. Our full strength was 16, and the average attendance of our first twelve meetings was between eleven and twelve. The attendance would have been even higher, but for the fact that many of our members worked late and frequently overtime.

We met in a warm room, were comfortably seated, and the set was placed in such a position that everyone could view it naturally. Most of us took notes during the talks and at least half of our number did some reading on the subjects. Often the talks dealt with subjects on which we had no expert knowledge, but occasionally we were privileged to have the company of persons with first-hand knowledge. Some of our members had travelled, others had hobbies of great interest, while many could speak with authority drawn from experience on the topics presented. It is a fact that after a year's membership of our discussion group, one of our number who was previously very shy and reserved, suddenly revealed unexpected ability as a speaker and gained confidence to such an extent, that she now occupies a leading position in our Dramatic and Debating Societies. In the English class at our Evening Institute, a young group member of 18 wrote his impressions of our group and the characteristics and views of members. When this was presented at our final meeting it caused great fun.

The value of the group lay in the fact that our discussions had more point than ordinary conversation; they did not stray far from the subject. Representative of various walks of life, our opinions had the spice of variety. We felt that by listening to the broadcast, and adapting it to our local circumstances, we derived help in solving our own problems and were enabled to understand those of people living in other places and in different and similar walks of life to our own.

A Librarian's Testimony

Two groups were organized, one to follow Professor Leonard Russell's talks on "The Modern Outlook" and one to follow Mrs. Wootton's "Some Modern Utopias." The average attendance of the first group was 27 and of the second 35. In connection with the first group 108 books were issued, and in connection with the second, 152. The interest aroused by these discussions was shown by the attendance of 120 people who assembled to hear a debate on April 6th on "Is a Utopia Possible?" The meeting lasted nearly two hours and a resolution was passed asking the Committee to continue the talks in the winter. This variety of persons and interests blended together splendidly in the discussions, and as the meetings progressed there was a real interest in the reflection of one personality upon another in the group. This might be taken as one of the distinctive features of the group meetings, and was a solid illustration of how an assembly composed of a variety of personality, belief, opinion and outlook, could blend together and discuss questions of importance without any distinctions of bias or feeling. The spirit of friendliness and comradeship grew to such an extent that there was a feeling of real regret, expressed and clearly shown, when the series of meetings came to an end. I personally witnessed the remarkable development of a working boy who at the first meeting in February expressed himself somewhat haltingly, but with a great deal of shrewd reasoning power, and who at the debate on 16th April was able to clothe his ideas in admirable English. Perhaps one of the chief values of these meetings lay in the feeling engendered amongst the members that the Library was indeed a "Public House" for the free, open and sympathetic exchange of views. There is no doubt of the stimulation of reading caused by these talks, as many of the books are still on loan and members have followed by-paths of the subjects through the library stocks.

The Experience of Some Unskilled Workers

The history of the listening groups which received the series of talks "Making Work Worth While" differs so much from that of the other groups in the Institute that it deserved special notice.

Two young production workers ("Machine-minder" type) at a large motor factory in the district had attended the Central Technical College during two sessions to study mathematics and drawing in order to fit themselves for promotion. Each session they had failed to "stand the pace" and had been compelled to give up. Last September the Principal of the College suggested that they attend the Evening

Institute where more individual attention might be possible. Both were backward and slow—but both were anxious to "get on." They made slow but steady progress whilst the classes held together. Then, due to various causes, the classes were closed.

The following week these young men came to talk over the difficulties the "machine-minder" experienced in the large factories and to ask for suggestions as to subjects of study to fit them for promotion *if it ever came.* Both felt that they had "rotten jobs"—jobs that "any fool could do to satisfy the boss (the foreman)"—and both felt that intelligence and ambition were useless in the productions shop. They had lost interest and were "fed-up" with their work!

The next evening one of them (22 years of age) came again with the request that assistance be given to form a discussion group for young "production workers." The series of "Talks" by Professor Pear on "Making Work Worth While" was to begin the next Tuesday. This man *suggested* that his group should be formed round these "Talks." The difficulties were discussed and finally he was offered the use of room and apparatus and told to "carry-on."

The Tuesday came, and Walker and his friend Austin arrived an hour before the "Talk" to see that all was prepared. They provided scribbling pads and pencils for note taking, and the Head Teacher had a copy of the *Aids to Study* Pamphlet and the *Listener* for each of the young men who were expected. Copies of as many of the suggested books as were available in the city libraries had been obtained and were ready for reference and inspection. (These were returned next day and were available for issue from the local branch library.)

Fifteen young men, all "unskilled" workers in the engineering industry, came in. The set was switched on, and at the request of Walker they were left in the room alone.

At 9:30 they were still engaged in discussion. At 10:00 they had to be asked to postpone the discussion until the next week.

The following Friday Walker came again to talk over his plans. "May they smoke?" "Do you think you could get a foreman from another works to attend the group?" "Can a list of easy books on the subject be provided?" Yes, these would all be attended to!

The Group continued with regular attendance of 18 members. The discussion was enthusiastic and general. The depth to which they delved and the amount of reading done were surprising.

After a month, the library loan of suggested books was withdrawn. Study and reading waned therefore. After one of the "Talks" the group asked for additional information respecting the Vocational Guidance and selection tests mentioned by Professor Pear. An Assistant Organiser of the Juvenile Employment and Welfare Department was approached, and a demonstration of the apparatus and a short lecture was arranged. But this must not interfere with the "Talks."

The series was concluded and the lecture was given. But this was not the end of it. Not a week has passed without a visit from one or more of the group to ask for reading or study guidance, or to put up some new problem. They have had meetings in the homes of the members, but they wish to come back to the class room next winter. . . . "It's too casual, and we ramble from the subject when we're at home." But, best of all, they say that they find the same job more interesting than before, and that the "unskilled" job is really skilled.

A Sample of Organization

From a numerical point of view the class was a failure, but to the small nucleus who attended regularly it was a success, and the discussions were most interesting and stimulating. The total number on the roll, including those who attended once only, was thirteen. Average attendance was 4.9. The members were:

1 Bank Manager	1 Student
1 Station Master	4 Household duties
1 Independent	1 Business Man
4 Clerical Workers	

The Vicar of the Parish was the Group Leader. The class assembled twenty minutes before the broadcast and talked over the previous week's broadcast. Afterwards individual opinions were called for, followed by discussion and argument, and if possible a conclusion arrived at. We experienced no difficulties in organisation—our only worry was lack of numbers, due to the usual apathy met with in this district when anything new is started. This was not due to lack of effort, for

(a) A dozen big posters made in school were posted in the area;
(b) 600-800 circular notices were delivered to houses and individuals in the area giving particulars of all classes running in the Evening Institute;
(c) About 80 special circular letters were sent to individuals whom we thought might be interested in the talks and might attend;
(d) Endeavours were made by personal contact to get others to attend.

A Rural Group

There were eleven members in the group and their trades and professions were:

Priest, clerk, roadman, retired Excise Officer, farm worker, bricklayer, married women.

When I first organised this group some years ago I made personal contact with people who were likely to be interested, and as the time passed those leaving the village have had their places filled by incoming residents and villagers whose interests have been aroused. The greatest difficulty was being able to fix a day. It was not possible to listen in to talks that would have met with more appeal, as there were so many other activities in the village, such as A.R.P. lectures, Red Cross and Evening Institute classes, which had to be fitted in to suit instructors coming from a distance. The discussion that arose showed that those in more humble walks of life had been about the world, and their experiences were invaluable. One man, now a farm worker, had been a miner in England and a foreman of native labour in Australia.

Those who are regular borrowers from the County Library were glad to avail themselves of the books sent by the County Librarian, whilst others did not read quite so widely. It was generally agreed that the evening spent was pleasant from a social point of view as well as being informative, and meeting at the school rather than at a private house prevented social distinctions from arising. No people gave up attendance because they did not enjoy the talks.

Groups as a Stimulus to "Good Mixing"

It serves a useful purpose in bringing together people from various walks of life and various occupations in a social capacity, to their mutual benefit by the free exchange of opinions. During these two years, although there have been differences of opinion, there has not been the slightest vestige of ill-feeling or personalities of any kind. There is a fine spirit which encourages even the youngest or most ill-informed to take part in the discussion with freedom, if he has something to contribute.

The Virtue of Informality

I have had listening groups for many years and I find it a great advantage to have these talks. It gives men of various ages that freedom for discussion you do not get in a formally organised debate. We do not allow any member to stand when he is speaking to the group—all remain seated and apart from the Leader who must keep the talk from wandering "round the Universe," there is no formally elected chairman or leader.

I find that the ordinary man likes the *informal* meeting where he can "chip" in as he wishes and does not fear being called to order.

Rubbing off the Edges

I find that the Group provides a healthy discharge for repressed extremist opinions. There is a salutary broadening of the mind when extremist meets extremist.

Was It a Failure?

In this town of 2,000 inhabitants, it was found there was too much competition among the numerous evening activities for the group to have a big membership. In the town there are a church, three chapels, a working man's club, the Junior Imperial League, Girl Guides and Boy Scouts and a territorial drill hall. During the winter each of these had some form of recreation or educational activity on at least one night a week. At least two of these bodies had debating societies, but they had no desire to co-operate with another listening group as they wished to keep their own distinctive denominational character. Seven members were quite sufficient for a good group, but as all except the housewives were active members of other associations, they could not attend regularly, and when it was found that on some nights there were only three present, it was decided to discontinue the group.

No reading was done, although the County Library offered to co-operate. A supply of books, however, came as the class was closing.

The group appeared to enjoy the discussions immensely. They felt they were taking part in something really worth while, and did their best to win recruits on this basis. They were glad of the opportunity to express their views, feeling it was good for themselves mentally, and good citizenship to be thinking about these matters.

A Sample Failure

We tried two groups—"Class" and "The Mediterranean," but in neither case was there enthusiasm or real life. "Class" had as its leader a man well-known in the town and for some time leader of a successful debating society; the invitations we sent to the men's political and other clubs gave no result and we gradually sank down to four members; then we closed. The History and Classics man of my staff [4] tackled the "Mediterranean" series, but the local League of Nations Union branch had no longer its enthusiastic secretary, and no response came from that quarter. I followed the first two courses as far as they went, and found them interesting, but not provocative, and therein lies the difficulty. There did not seem to be sufficient unpremeditated give and take between the speakers; they were too friendly to one another, and they did not leave our groups itching to join in. The suggesting of formal questions for discussion at the end of the half-hour left the group wondering rather than stimulated to expression.

A Semi-Permanent Group

The Leeds Wood-house Moor Library group has devised an interesting way of keeping the group together during the summer. It was considered that once a week was too much for the group after

[4] This witness is a secondary school Head-master.

attending regularly all the winter, and therefore they meet fortnightly on Thursday evenings to discuss a topic which one of the members introduces in a talk of fifteen to thirty minutes. Occasionally radio talks from the general programme are used, but on the whole it is found that they are unsuitable in time for the purpose of this group. There was an average attendance of thirteen throughout the summer. This group has been one of the regulars for many years, and incidentally has the charming custom of adjourning to the pub across the road at 9:30 P.M. when the discussion is continued over bread and cheese and beer.

Can the Listening Group Become a Class? [5]

At present opinions vary concerning the values of group listening. It is generally conceded that the wireless discussion group makes some original contributions to informal adult education but its possibilities of turning a group into a regular adult education class are undetermined and debated. One of the most recent pieces of evidence (*Group Listening in Gloucestershire,* a report of an experiment conducted by the Education Committee of the Gloucestershire Rural Community Council, session 1938-39) dealing with the problem of standards of achievement where group listening is concerned, believes that

Where talks have been used for Listening Groups composed of people not familiar with adult education methods, most emphasis must be placed upon the discipline of regular meetings and the value of self-expression; the educational value is not confined to the information given in the talks. Listening Groups as a supplementary activity to other forms of adult education, however, may very well, through the participation of experts and the stimulating methods of presentation which can be adopted, add significantly to the work of lecturers and tutors.

We decided to experiment with Eric Newton's "Artist in the Witness Box" series, broadcast in the winter period of 1939 and Spring of 1940 to find out how far "class-wise," without destroying the informality integral in successful group listening, it was possible to go. We knew that our group of fourteen members, each of whom paid a half-crown fee to the Education Authority for the course, could not be reckoned initially as "students."

[5] This section is contributed by Mr. J. H. Higginson, Assistant Director of Education for Lancashire, in collaboration with Mrs. Higginson.

They included a girl in the early twenties, one or two young mothers, and the rest were middle-aged men and women. Their interest in and knowledge of pictures was that of the ordinary person "of average intelligence and average education": one member in her youth had attended an art school, two had been abroad and visited the Louvre and seen other well-known collections of pictures, and these people seemed to have been trained to admire certain pictures, e.g., the *Mona Lisa, The Laughing Cavalier, The Doctor, Ramsgate Sands.*

The atmosphere of the first meeting was charged with the thoughts "I'm going to have my leg pulled and I shall resent it"; "I'm on the defensive where art is talked about nowadays"; "I've not forgotten Epstein's *Adam* yet". This uneasy atmosphere prevailed throughout the talk and well into the discussion. The first remarks: "I'm convinced modern artists are just out to make a name and try to do it by shocking the public, like Epstein, or by pulling its leg: you'll never make me believe they're sincerely trying to depict beauty" and "I hope you're not going to expect us to like modern stuff"—show how nervous and sensitive the group were in the beginning. Some members said they were not "arty" and knew nothing about it, but they liked pictures. After the release of such comments we discussed the broadcast a little and decided to set out with Hodge in the spirit of "I don't understand modern art, but I hate to feel I'm missing something." As we had not received the illustrated pamphlets from the Publications Department we looked in a tentative way at about thirty reproductions ranging from Mexican prehistoric art to the art of our day which was represented only by the more easily comprehended pictures.

By the second meeting the group had B.B.C. pamphlets and a box of books from the County Library. This Library has a well-stocked fine art section: plenty of standard and recent books and a fine set of reproductions including contemporary work. The books for the box were chosen carefully with the first meeting well in mind: books of reproductions, others in which reproductions predominated and reading matter merely commented, a few biographies, and books which examined specific periods in detail, e.g., *Reflections on British Painting* by Roger Fry. Following Eric

Newton's guidance in the broadcasts the books were "calculated rather to stimulate taste than to form it, to excite curiosity rather than to satisfy it."

As the weeks passed an easier atmosphere succeeded. Every member was regular in attendance, keen and reading, but even at the most enthusiastic moments one felt enemy prejudice was not dead, only lying low.

The group met in the parlour of a Free Church as this was the only suitable and central room in the locality. Two members acted as secretary and librarian, both interested and efficient. The librarian arranged her bookbox and table near the door and as members arrived they changed their books. Then as a rule they moved to the fire and, as they chatted and warmed themselves, looked at the reproductions on the mantelpiece which each week were designed to help to prepare their minds for the broadcast. About five minutes before the broadcast we read through the week's pamphlet notes stressing any questions raised and any specific terms, as Academic, Abstract, and Representational Painting. Several members took notes and after the broadcasts points which had not been understood or agreed with were brought up and discussed.

The group were familiar with little more than a few Italian madonnas and nineteenth-century naturalistic art. To remedy this it seemed necessary to get the group to look at quantities of pictures and each week we had a picture show. After Talk IV ("The Artist and His Race"), we examined twenty pictures to reinforce the talk and each member answered a short questionnaire on two or three pictures. Then the pictures were held before the group and the answers given: the results were instructive, stimulating, and sometimes very amusing. The procedure was liked and we tried variations on it. Talk VI dealt with abstract painting, of which the group had frequently expressed dislike: the picture show consisted of reproductions of different types of nineteenth-century paintings with which they were familiar, "flat pattern" and "mountain-of-bricks" cubism, war pictures of Nevinson, William Roberts, La Fresnaye, and Wyndham Lewis, and more recent works of artists who show the effects of cubist discipline— Paul Nash, Wadsworth, Picasso, and Mrs. Dod Proctor. As we explored the pictures we tried to discover (a) answers to the questions in the pamphlet notes on Talk VI, (b) what in nineteenth-century art, contemporary artists have revolted against, (c) if the picture enlarged our experience in any way. After this show, enemy prejudice appeared to have been dealt a heavy blow: it was the first time some members had looked with real curiosity at difficult-to-comprehend modern pictures. Also each member took home one modern picture to live with for a week bearing in mind the question, Is this picture enlarging my experience in any way? The following week the pictures were returned with comments: these were varied—a man had found it exceedingly difficult to live with one of Rouault's ladies and did not wish to repeat the experience; the following week he took a Matthew Smith landscape and found it much pleasanter. This experiment was also repeated at the group's request and it may have given "time" a chance. Another week we had a one-man show, the group leader gave a short account of Picasso and we examined pictures covering all his periods.

Most of the broadcasters made good use of the pamphlet and by the end of the series the group were so familiar with it that Mr. Newton's "most powerful advocate time" had had a chance. In the last meeting we went through the pamphlet trying to remember first reactions to the plates and to assess our feelings after twelve weeks' familiarity with, discussion and controversy over, them.

Members who had not time or inclination to read through books were eager to read recommended odd chapters or sections of chapters, or articles from *The Listener;* these latter which had been collected over a period of years made a valuable addition to the bookbox. Discussion about books read was encouraged; sometimes a member would ask if he might read a paragraph which he felt would be of general interest and in this way we also became familiar with the art views of the *Daily Mail* and *The Observer.* From time to time the group leader drew attention to certain books; for example, when Plate 42 (Stanley Spencer's *The Resurrection: Burghclere Memorial Chapel*) caused a good deal of difficulty, Appendix V in Wilenski's *English Painting* was recommended.

We feel that the "Artist in the Witness Box" has permanently enlarged not only our attitude to pictures but our realization of other things: the mothers were particularly interested when we compared primitive art with the art of children, and at Christmas one brought her seven-year-old son's term's painting to show us. Another member who before the course had not heard of Cézanne is now keen to see the red earth of southern France for himself. The ground seems prepared for seed: the group are eager to visit exhibitions and see new pictures, revisit galleries and test old favourites by newly awakened interest, and some of the younger members are keen to try creation themselves.

Opportunities of visiting exhibitions of pictures are rare in this locality and the group are now looking forward to a small exhibition which is being arranged in connexion with the annual rally of groups in April when a lecture on "Ways of Looking at Pictures" will be given. In retrospect we feel that the members of this group have achieved more than the discipline of weekly meetings and desultory comment even though the latter may have in it some element of self-expression. The path from prejudice to reasoned conviction cannot be followed to its end in a twelve-week course, but the group developed from mere casualness of interest to a seriousness of purpose worthy of comparison with that of a good adult educational class of the more elementary type.

Summing Up

So many of the "consumers" quoted in the foregoing pages express such a favourable opinion of group listening that a final word of caution is called for after considering their evidence, especially as what they say may seem to rebut some of the general criticisms expressed in other chapters of the Report. These witnesses leave no doubt that group listening is often a most exhilarating and illuminating experience. On the other hand the people who have been disappointed with the experience are far less willing to respond to an invitation to give their views. They are the ones who have drifted away and are hard to locate. But they exist in at least as large a number as those who praise group listening. It may be as well, therefore, to reiterate the main criticisms which group listening has evoked from many competent observers in Great Britain:

(a) Although the value of group listening is considerable, it remains a lesser force in adult education than some zealots maintain. It is one of a bunch of informal and elementary activities which, so long as too much is not asked of them, can provide an effective approach to more serious modes of study.

(b) The extent and growth of group listening must always be considered in relation to the massive organization and impressive expenditure by which this movement has been sustained by the B.B.C. If several other movements in adult education had been given such expensive and continuous blood transfusions their rate of progress would certainly have been more notable. So long as these two considerations are appreciated, group listening can be assessed in its true proportions.

This chapter may well end with a balanced and good-humoured commentary by one of the most experienced observers of British groups. Here it is:

I have lifted the latch of village schools, climbed up the hill to the unemployed club, given away the marrow at the Women's Institute, sat on the radiator in the public library, sipped the too-sweet tea in the suburban sitting-room, stumbled down the pitch dark road to the British Legion hut, fought through the rain to the Y.M.C.A., in fact seen good Groups and bad Groups. Some were merely sounding-boards for the vanity of the leader, some were immature political meetings, some were merely chatter, some indescribably dreary, some a whole Golgotha of King Charles' Skulls. And some, by no means a few or even a minority, were all that the eulogists at the B.B.C. claim for Groups. To attend one of these last was a really encouraging experience, not because the work approached Tutorial Class standards, but because men who were getting interested in art or politics or what-not were enjoying conversation about something they felt mattered. It is always difficult for those who get full opportunity for the exchange of views in the ordinary course of their lives to appreciate the part played by something like a Discussion Group in the lives of those who don't.

V ⋅ GROUP LISTENING IN OTHER
EUROPEAN COUNTRIES

IN EUROPE the leadership in group listening unquestionably
rests with Great Britain and is long likely to do so. So long as
totalitarian régimes last in Germany, Russia, and Italy, discussion
groups will be on the long catalogue of prohibited activities.
France, the most insular country outside the U.S.A., is not likely
to change its mind about the uses of wireless, and will continue
to reject wireless discussion as a canned substitute for the real
thing. The French, the most persistent and effective debaters in
the world, will no more accept wireless discussion than they will
accept tinned soups as a substitute for their celebrated *potage*.
The Balkans are still in that rudimentary stage of cultural development at which the wireless cannot yet take a hand in the
selective manner of talks for discussion. There remain half a dozen
European countries where wireless education for adults in one
form or another has got a foothold. In Sweden, Norway, Finland,
and Czecho-Slovakia the system of wireless education includes
some development of group listening; and some account of its
progress in these countries forms the substance of the first half of
this chapter.

The summary is for the most part factual, since the values and
problems of group listening as a whole are considered elsewhere
in this Report. Information about the actual number of groups
existing in any of these countries is difficult to secure with any
exactness; the aggregate of group listeners is obviously even more
difficult to check. But not even in Sweden or Norway would the
claim be made of more than four or five thousand group listeners
for any single series.

Sweden

After Great Britain, Sweden seems more anxious to develop group listening than any other European country; but although attempts have been made since 1932 to get groups on a solid basis the achievement has been disappointing. The Swedish Broadcasting Company made itself responsible for organizing groups at first, but in 1936 decided to hand over this responsibility to the adult education movement, while retaining for itself the responsibility for producing the talks. The immediate result was a notable decline in the number of groups—a result which bears out the experience elsewhere—that unless a Broadcasting Service carries the baby, group listening comes to grief. Nowhere does organized adult education seem disposed to rate group listening as an activity of special value. In Sweden, there appears to be in many districts keen rivalry and competition among the various adult educational bodies, and the effort to attract students to this or that educational activity is intense. None of these bodies is convinced that group listening would assist them in recruitment, and none of them has been persuaded that, by taking it up seriously they might get ahead of their competitors. Their general attitude to group listening, therefore, ranges from tepid interest to utter indifference. In Sweden, to a far greater extent even than in Great Britain, group listening has been kept alive only by the careful nursing of the Swedish Broadcasting Company. The Swedish broadcasting authorities, however, are not in the least inclined to abandon the attempt to secure a fuller co-operation from the adult education organizations, and it has lately appointed a special officer to foster group listening among these organizations.

Despite these difficulties there has been much effective group listening in Sweden. Groups have done best in rural areas, where urban distractions are absent and where the villages are too scattered to enable more ambitious forms of adult education to be attempted. Language courses in English, French, and German are as popular in Sweden as in Denmark, with the difference that Sweden, unlike Denmark, has been able to some extent to put foreign languages over to groups as well as to fireside listeners. Much has been done to develop the efficiency of these language courses, which are based upon specially written graduated text-books which command a sale running into many thousands. For instance, a control group has been assembled at the headquarters of the Broadcasting Company so that the broadcaster can check up on the extent to which he is giving his wider audience what they need. The control group listens to each broadcast, after which the wireless tutor comes to teach them and to discover what they have absorbed and what they have missed. Some idea of the keenness there is for these courses may be gained from the fact that for the first such control group there were 600 volunteers for 10 places. A conservative estimate of the number of regular listeners to the language courses is 50,000, but a minority of these are found in listening groups.

Elaborate work has also been done in another special province of Swedish popular interest—drama. The procedure is, first, to publish as a pamphlet a "classical" play—Sophocles, Shakespeare, Shaw (all in Swedish translations). At the same time the Swedish Broadcasting Company issues a "study-letter" on the selected play. Subsequently the play is read on the air, so that by the time this triple process of preparation has been completed the groups are well equipped to discuss the play. Nowhere else in Europe has such a scheme been attempted, but in Sweden it has met with remarkable success.

The Swedish broadcasting authorities are as convinced as the British that the quality of group leadership is a decisive factor in the organization of a continuous system of group listening. Sweden has not hitherto provided any facilities for training leaders, but from time to time special conferences of adult educational bodies are called for the purpose of expounding forthcoming series of talks and discussing the most effective ways of considering the topics in the groups. Sweden has not so far experimented with paid group leaders, and the authorities are not in agreement about the wisdom of such an attempt. But one of the directors of the Swedish Broadcasting Company is a thorough-going advocate of paid group leaders, and, regarding it as a branch of the teaching profession, wishes to see it subsidized by the Government.

In its methods of presenting material for groups, Sweden has done little more than the "straight" talk: it has seldom attempted microphone discussions, on the British model, for example. The reason advanced for this lack of experimentation is that the microphone discussion demands more rehearsal than they can afford. But British experience certainly confutes such a view: the microphone discussion—based on a script prepared by the participants —calls for no more rehearsal than the "straight" talk. On the other hand, Sweden has experimented with some success in a collaboration between its Talks Department and its Features Department. The expert, under the aegis of the Talks Department, provides (so to speak) the letterpress of the talk, while the Features Department fits in the illustrations. This device, however, appears to have appealed less to listening groups than to the individual listener, and for the most part the groups are now provided simply with the "straight" talk.

One of the best features of the Swedish group listening system is its service of supplementary publications. Apart from their well-annotated programme pamphlet they issue an admirable pamphlet for each series of talks; and for many of the series they supply a weekly "study-letter" which annotates the broadcast material. Moreover, they publish a series of text-books based on talks —and this series appear to be a remarkable success. Their handbook on housing, for instance, sold 10,000 copies at 2/ each; and this, like many of their best-sellers, is such a complete work of reference as to retain its selling power long after the talks are over.

The favourite topics for group listening follow the same general lines of interest as in Great Britain, especially in international affairs, economics, and government. But Sweden appears to be more interested in problems of child welfare, on which topic there have been four big series in the last four years. Their appetite for history, too, is more notable than that of the British, although it is admitted that a topic such as the "French Revolution" (in its 1940 programme) is likely to be more informative than debatable.

On the whole, then, Swedish broadcasting is favourable to group listening, and, despite discouragement, has persisted in the attempt to get group listening on its feet. So far the movement has been intermittent rather than systematic; and the broadcasting authorities have come to believe that its expansion will depend —as it did in Great Britain—on the amount of time and money the Broadcasting Company can spend on fostering the use of group listening among the adult education organizations. Those bodies, as in Britain, will need a good deal of prodding before their lukewarm attitude to radio education is transformed into real collaboration. The Swedes seem determined to persist in prodding these reluctant partners; and had it not been for the alarums in Europe during 1940, they would have carried the process a good deal further. As it is, Sweden can rank as the second-best of the European countries in its development of group listening.

Czecho-Slovakia

The title of runner-up to Great Britain would have gone to Czecho-Slovakia rather than to Sweden—until 1938. Until Czecho-Slovakia became a German province it had carried group listening a very long way. The numerous educational and cultural organizations of the country had been very willing to take the special series organized for them by the Czecho-Slovakian Broadcasting Company; and among the most prominent of these collaborators were the Masaryk Institute of Popular Education, the Sokols, the Parent-Teacher Association, the Child Hygiene Society. Their part in providing the audiences was not, however, the major one; for the Local Education Committees (i.e., the State System), largely their schoolmasters, usually took the lead in convening the groups.

The period devoted to series for groups was given a fixed place in the programme—every Friday evening from October to March; and there is no doubt that the insistence on a fixed period, which soon became universally known, had much to do with the rapid growth of group listening. The Czechs found that a series of four was better than the British practice of a series of twelve; but the difference indicates that the Czech groups were really much more elementary than the British. This point is of some significance when we are faced with the abnormally large group listening audience which Czecho-Slovakia registered.

The Czechs showed a lively readiness to experiment with

methods of presentation. For instance, they broadcast imaginary meetings of child guidance societies, so that listeners could get a vivid idea of what such specialists dealt with. Their series, again, often had a strikingly practical objective, so that listeners were stimulated to act as well as talk. The sponsors claimed to have instigated the acquisition of playing fields, the organization of Fire Brigades, the establishment of a communal dispensary, and so on. One famous series was the well-known "What Can You Do For Your District?", which is said to have had a remarkable effect in tuning up popular interest in local government—in town-planning, hygiene, agricultural development, education. This series attracted to its course of four talks on town-planning the astonishing total of 4,128 groups with an aggregate group membership of 173,000 people: figures which have not been approached anywhere else in Europe. When we consider, however, that these figures work out at an average number of more than forty people to a group, it becomes evident that they were not discussion groups in the usual sense. British experience shows that effective discussion (as distinct from comment or question) is possible only in groups which range from a dozen to a score of participants. Many of the Czech groups were, in effect, small town meetings, with the Mayor presiding, followed by a few questions and one or two supplementary addresses. At the same time, it would not be fair to suggest that this big audience has merely a casual and passing interest in the course, for this series on town planning produced no less than 5,000 letters from group listeners.

The Czechs appear to have secured a far greater degree of collaboration from educational bodies than any other European country has managed to command; and when the freedom of the Czechs was filched from them, they were demonstrating once again their zeal for education by this rapid development of group listening. There can be little doubt that their group listening movement would have been consolidated into a most valuable system of civic education in a few years' time. One report, furnished a short time before the Germans marched in, ran thus: "In this town of 11,400 people, the organizer has built up 29 Groups comprising more than 3,000 persons in all." In several country districts, the schools and restaurants where the groups met were so crowded that listeners stood in the corridors or under the windows. All this may not be group listening of the precise and systematized kind which has been developed in Great Britain or Sweden; but it is evidently the kind of popular interest from which might have emerged a system of group listening second to none. Apart from that, no European country had been more quick than Czecho-Slovakia to respond to wireless as an educational system.

Norway

Group listening in Norway dates from the period when broadcasting was nationalized in 1934. At first it was a lively success, and in its first year attained a total of 1,000 groups for a series of talks on "The Labour Movement." The groups are not directly organized by the Broadcasting Organization but by such well-known Norwegian cultural and social organizations as the Workers' Educational Association, the Young Farmers' Movement, the Labour League of Youth, the National Farmers' Union, the Temperance Movement. These bodies remain, in the words of the Norwegian Broadcasting Organization "the dorsal fin of our Group Listening movement." Many of the Norwegian groups have as large a membership as 40 or 50, but the authorities do not conceal their view that such large groups do less effective work than the small group of 8 or 10. The series, as a rule, comprise no more than four or six talks—in contrast to the British average of twelve talks.

The Norwegian groups, like the Swedish, are well served with study-letters, explanatory pamphlets and similar preparatory and documentary matter. Many of the most popular series of talks have subsequently been published in book form and continue to be in demand many years later. They particularly encourage the use of public and institutional libraries by group members—an emphasis which is sometimes neglected in this work.

The most popular topics have proved to be the customary solid choice of adult education—international co-operation, local government, "educational psychology," the position of women, the social services, the Norwegian legislative system. A few years ago one innovation—"The Internal Combustion Engine"—proved a striking success, and was followed up by an equally "practical" course on "Building Construction," which proved no less popular. No other European country appears to have experimented in this blend of adult education and technical education.

The Norwegians have not inaugurated any system of training for group leaders, but they are convinced of the critical importance of such training courses, and they have at least gone so far as to convene conferences of group leaders to discuss the problems of their job.

Finland

This cultured and progressive little country has been as keen as Sweden in the attempt to build up a system of group listening. The Broadcasting Organization did not for some time attempt to organize groups itself, but included in its programmes special series for the study circles which form such a large part of the Finnish system of adult education. The three main providing bodies have, between them, something like 3,000 such study circles, and these have been well served by the Broadcasting Organization. The special transmission for these existing study circles takes place on Sunday afternoons; and they are reinforced by pamphlets on each series. Outstanding topics in the past few years have been "The History of Finland," "Temperance Reform" (always a popular topic among the Finns), "Local Government," "Finnish Heroic Poetry."

The main purpose of the Finnish Broadcasting Organization, then, has been confined to a supplementary service, in the fullest sense, to the work of the existing democratic cultural organizations. It has not sought a more ambitious line of development, and possibly for that reason has at no time been suspect by those organizations. It has been content to serve their existing cultural interests rather than lead them into new ones; and although the scope of group listening in Finland is thus a deliberately restricted one, the service it provides has been more acceptable to adult education as a whole than it has managed to be in any other European country.

The point was made earlier in this Report that the absence of group listening in certain countries does not necessarily mean an indifference in those countries to wireless as an educational medium. The remainder of this chapter will note briefly some European examples of countries which, despite their notable provision of educational talks, have either not yet developed any system of listening groups at all, or else have done little more than toy with the idea.

Holland

Radio programmes in Holland are supplied by four separate services: a Protestant system, a Roman Catholic system, a labour system, and a general undenominational and non-party system (AVRO). All four systems provide a very reasonable ration of educational talks, yet none of them is in the least disposed even to experiment with discussion groups. They consider that the proper use of wireless talks is to transmit information, and even when talks are received by a group of club members they are received in an assimilative silence. The themes of these educational talks are for the most part solid and unprovocative: academic pleas for an international understanding; dissertations on the first principles of Christianity; chapters of Dutch history. The most popular educational topic of all appears to be astronomy, and there seems to be no limit to the Dutch capacity to be informed about the motions of the planets.

Pressed for an explanation of their indifference to group listening, the Dutch broadcasting authorities asserted that adult education is so well developed in Holland as to provide a maximum measure of opportunities for discussion, and that the broadcasting services therefore saw no point in seeking to extend such opportunities—at least until the adult education movement demanded such a reinforcement.

One recent development has in it a possibility of becoming a type of group listening—the wireless programmes of the Radio-folks Universiteit (RVU). This is a service (set up in 1931) whereby a small number of short lecture courses are transmitted on a loaned wave-length, half an hour a week, to listeners at their own firesides. Although this system calls itself a radio university, it reveals only the most elementary form of university teaching, since it lacks all discussion opportunities. The system has a following of some 16,000, whose subscription to the weekly journal pays for the whole set-up. There is evidence that these programmes often instigate family discussion and discussion among groups of friends. But so far no attempt has been made to organize group listening at the reception end. Typical themes of this service are "Man and the Universe," "Goethe," "Human Society," "South Africa," "Nutrition," "The Intelligence of Animals."

Switzerland

Swiss broadcasting is complicated by its triplicate system of transmitting in French, Italian, and German—a system which makes overhead costs and running expenses so high as to impede many possibilities of experiment and expansion. Italian listeners in Switzerland, for example, who number 15,000, would have no service at all were it not that their programmes are subsidized by the French and German-speaking majority. Although the original seven independent and private broadcasting systems have now been combined in a central administration, the seven original stations are still in use, and this kind of complication continues to handicap the development of a unified system.

The Swiss authorities agree in their high rating of the educational purposes of broadcasting, and all the programmes contain a very reasonable "dosage" of serious talks. The German station, for example, gives half-hour educational talks each evening, and resists a majority of its listeners who complain that the educational emphasis is being overdone.

An organization called the "Reader's Association," under the direction of Paul Lang, has made attempts to organize groups. The attitude of the Swiss broadcasting authorities has been: "If you organize them, we will furnish them not only with a programme, but with text-books, etc., on the lines of Sweden." This attempt has in effect been completely unsuccessful; there is evidently no organic demand for group listening. All witnesses declare that the Swiss temperament is against discussion and Swiss educational systems do not train them for it. At the universities, for example, there are no debating societies among the students, and the tradition seems to be well grounded that education is a process of listening and observing rather than of answering back. Even in the Swiss Parliament there is no debate: speakers simply read their piece from the rostrum and return to their place. This peculiarity was universally testified to by Swiss authorities, and it raises a disturbing thought about Education for Democracy and all that. A further factor raised by some witnesses is that political and religious difficulties in Switzerland are so acute that they do not like to disturb sleeping dogs by having discussion.

The Basle system broadcasts a certain number of special talks in the evening programme, talks of a decided educational emphasis, but there is no organization to see whether these talks are listened to. In the same way, they run courses in foreign languages, which they know to be popular, but they have no evidence as to how many people listen or whether any of them listen in groups. In the remote mountain villages there is often only one radio set to the community; in such cases there does exist a certain amount of communal listening, but no information can be secured about what happens at these accidental listening groups.

School broadcasting in Switzerland is extensive. School broadcasts are in some cases an integral part of courses of instruction at commercial and technical high schools. Part of the curriculum in these places is, in fact, given by broadcasting—a privilege which appears to be entrusted to broadcasting nowhere else in Europe. The talks given to the schools are frequently rebroadcast in the evenings for adult listeners—yet even this opportunity has not been used to organize listening groups.

Denmark

Denmark has as keen an educational emphasis in its broadcasting as Holland, and yet has been unable to establish a system of group listening. One of the reasons advanced is that adult education is so well developed as to leave no elbow room for auxiliary services; another is that adult education is so satisfied with its own methods as to look with little favour upon such a limited instrument as broadcasting.

There is an immense amount of private listening to serious talks; and, as in Sweden, there is a big demand for the foreign language courses, which are organized as a two-year course and which provide two lessons a week in each of three languages— English, French, and German. Besides these language lessons, the Danish radio broadcasts one talk a week each in those three languages.

Denmark, however, unlike Holland and Switzerland, has at one time or another experimented with listening groups. In 1933, a special series of talks on "Democracy and Dictatorship" was provided for the Women's International League for Peace and Freedom, but it appeared to attract no more than 400 listeners organized in as many as 50 groups. The following year a similar special series was arranged for the Mothers' Association, on the topic of "Heritage and Race." In 1935, at the request of a group of youth organizations, the Danish broadcasting service gave two series of talks on "Youth and the Land" and "Denmark's Foreign Trade." None of these special courses was sufficiently successful to encourage further development, and since that time nothing has been done. A new effort was planned for the autumn of 1939, but has been abandoned because of the war. But one point deserves to be particularly borne in mind: had the Danish broadcasting system been able to afford to persist in its effort to establish group listening, it might well have established a movement which would have developed. It has been nobody's business to look after the tender shoots planted in 1933, 1934, and 1935; and the Danish broadcasting authorities are inclined to believe that, if they had the means and the organization to keep up a protracted effort, they might well get group listening going in Denmark.

Australia

One of the youngest systems of group listening outside Great Britain is that inaugurated early in 1939 by the Australian Broadcasting Commission. It began in the State of Victoria when invitations to form groups were broadcast direct to listeners from one of the State stations. In this way, no fewer than 70 groups were formed to listen to a series of 12 talks. More than half these groups met in private houses; and the inquiry conducted at the end of this experimental scheme showed that the listeners were

on the whole a constituency different from that of adult education in Victoria. Several of the groups consisted of people living
in isolated country districts, many of them quite illiterate. For
several reasons some of the groups do not meet for discussion on
the actual night of the broadcast; the members listen in their own
homes and meet together on the following Sunday to discuss the
talk. That improvization has certain serious limitations, for in
the interval between hearing the talk and discussing it many
points of the argument are lost or blunted. This disability was
to some extent mitigated by circulating copies of the talk among
members of the group.

Following this move in Victoria, the Australian Broadcasting
Commission then decided to appoint an organizer of listening
groups for the State of New South Wales, and in the autumn of
1939 he began work, in conjunction with a series on "Burning
Questions"—"Slum Clearance" (3 talks); "Defence *versus* Social
Services" (3 talks); "Australia and the Refugees" (5 talks). This
New South Wales scheme, like the Victoria one, was organized
directly by the Australian Broadcasting Commission; but this
time in close and active association with the Workers' Educational
Association, the University Extension Boards and the State Departments of Education. Among other bodies who were asked
to arrange groups among their numbers was the Australian Na-
tives' Association.

Much attention was paid in this second Australian experiment
to the problem of group leadership. University and W.E.A. lecturers visited several groups to demonstrate the best methods of
discussion; and the Australian Broadcasting Commission also
broadcast a short specimen talk, which was followed by a sample
discussion.

It is too early to say how far this Australian venture may
spread or what results it will achieve; but the authorities are well
pleased with the early interest shown in group listening, and are
particularly gratified with the willingness of adult education
bodies to collaborate. The Australian Broadcasting Commission
has been particularly well advised in appointing a full-time officer
so early in the scheme; for it is in this difficult stage that several
European countries have lost the chance of developing group lis-

tening because it was nobody's particular business to nurse it and coax it along.

Group listening has not yet begun in New Zealand; but the Director of the New Zealand Broadcasting Corporation who was formerly a university professor is projecting an experiment on the lines of the Australian schemes.

VI ⋅ SCHOOL BROADCASTING IN GREAT BRITAIN

Links with Group Listening

SCHOOL broadcasting, for two reasons particularly, is worthy of a chapter of its own in this Report. The first of these reasons is that in Great Britain the organization of group listening is closely linked with that of school broadcasting.[1] Although it may have been, in the first place, a mere matter of convenience to make the Chief Executive Officer on the school's side responsible also for the administration of group listening, that act of central co-ordination has produced some valuable developments. One of the most significant of these developments has been some experimentation in that No-Man's-Land of youth, that period between adolescence and adulthood which is so critical a constituency in any nation. The two partners in British group listening—the schools and adult education—have during the last year fostered a very successful new venture called the "Under Twenty Club." The broadcasts have been widely used in evening institutes and girls' and boys' clubs, and the programmes themselves are in the form of club meetings. Each week a number of young people, under twenty years of age, meet in the studio to discuss some topic of current interest. A notable feature of this series is that the discussions in the studio are extempore, although the ground has been prepared previously in an informal discussion among the participants. The subjects include sport, travel, work, leisure, politics, foreign affairs, citizenship, entertainment, careers, and so on. There is an adult chairman, and the club members themselves are drawn from all parts of the country and from all

[1] See the opening pages of Chapter IV.

classes. Sometimes the problem to be discussed is crystallized in a dramatic scene; sometimes an expert on the subject visits the club and is cross-examined by the members. The series is intended to interest young people under twenty, in particular those who are already earning their own living. The broadcasts are especially suitable for listening groups, since at the end of each meeting the chairman sums up the points which have been covered and indicates questions which have been left open and still require discussion. The young people at the microphone speak their own thoughts in their own words; it is the part of the chairman to train them in getting at facts, seeing other points of view, and thinking clearly without hampering the spontaneous expression of feeling and opinion. This novelty has introduced devices of production (e.g., the dramatic interludes) which adult groups have not yet dared to tackle; and its other device, of impromptu microphone discussion, has been far more successful than when it has been attempted for adult groups.

A visitor to one of these "Under Twenty Clubs" in Wolverhampton reports:

After the broadcast quite a number of these young people had something to say, and this perhaps was the most significant thing about the meeting. Boys and girls of that age are normally too shy to say anything at all. Very little of what was said had any real significance, but the headmaster was delighted that they were willing to give their views. He regarded the "Under Twenty Club" as an excellent means of achieving a difficult purpose, and said that he would certainly carry on with the club so long as these broadcasts were available. Actually the broadcasts took the place of the English lesson at this Evening Institute, and the headmaster's opinion was that the material provided and its power to evoke discussion was an excellent supplement to a formal lesson.

The "Under Twenty Club" is one of the war-time casualties of British broadcasting. Time simply cannot be found for it in the congested single programme which has replaced (for military reasons) the half dozen simultaneous programmes of pre-war days. In its few months of existence it suggested a valuable way of filling the gap between school broadcasting and group listening, and it is one of the programmes which will immediately win back its place when peace returns.

Is School Broadcasting a Nursery for Group Listening?

Some observers in Great Britain contend that school broadcasting is a valuable method of preparing girls and boys for group listening when they grow up. There is no evidence to show that any notable number of members of listening groups developed the habit as a result of listening to school broadcasts; that issue must remain a matter of speculation.

School broadcasting, if it is successful, should lead to more selective listening as the children grow through adolescence into responsible adulthood, but selective listening is not by any means the same as having a taste for discussion talks. The majority of people who are interested in good talks are still not interested in turning out of their houses in order to discuss them with a more or less arbitrary selection of neighbours. School broadcasting should perhaps make for better discussions within the family circle, but there is no *a priori* reason for supposing that it will lead to any great increase in the demand for further education of the listening group type. Sixth Form talks should stimulate adolescents to thought, but those who pass through Sixth Forms successfully are likely to pass beyond the level of radio discussion groups and to pursue their interests by the more satisfactory method of reading—except in one or two instances where the radio alone can do a job. For instance, international affairs can be treated more topically on the radio than in books, and topicality is of particular importance in this field. On the other hand, the peculiar responsibility of the B.B.C., built on its reputation for cautious honesty, often makes it difficult for it to be as frank in its speaker's analysis of foreign affairs as educationists would like it to be. Then, again, a series on art criticism is peculiarly suitable to radio, since the material is just not available in books. But apart from these specialized subjects it seems unlikely that better-educated school children will grow up into more discussion-group-loving adults. The "Under Twenty Club" is much more likely than school broadcasting to develop a later predilection for group listening.

History of School Broadcasting in Great Britain

Within a few months of its foundation, the B.B.C. appointed a Committee including representatives of local education authorities, directors of education and organizations of teachers to advise upon, and watch the progress of, educational broadcasting, It was found that a number of schools were anxious to experiment with broadcast talks specially planned to meet their needs, and in the summer term of 1924 the first experimental series was given, followed in September of that year by the introduction of a regular service. By June, 1926, the broadcasts had already so established themselves that it was felt possible to conduct a definite experiment into the use of school broadcasting, and with the aid of a grant from the Carnegie United Kingdom Trustees a year's experiment was conducted in Kent schools by the Kent Education Committee and the B.B.C. in co-operation.

In 1927, the British Broadcasting Company was replaced by the British Broadcasting Corporation. The next year marked the end of the first stage in the development of school broadcasting, for in the Kent Report published in that year forty teachers indicated that after a year's experiment they were not willing to be deprived of wireless as an aid to teaching. This view that school broadcasting was emerging from the purely experimental stage was shared by the original advisory committee appointed by the B.B.C., which voluntarily resigned in order to facilitate the establishment by the B.B.C. in 1929 of a Central Council for School Broadcasting, to guide the development of the school broadcasting service. Its function was, and still is, to advise the B.B.C. both on contemporary educational policy and on the detail of educational practice, and to secure the recognition of school broadcasting by teachers as an activity sponsored and recommended by a qualified educational body.

In 1935, the B.B.C. decided that the time had come when the Council should be granted a greater degree of independence. The essential purpose of the Central Council remains unchanged, but the greater independence granted in 1935 resulted in a division of functions between the B.B.C. and the Council which is briefly as follows:

The programmes broadcast to schools are produced by the Schools Department of the B.B.C., under the direction of the director of school broadcasts. The B.B.C. has an overriding power in respect of Corporation policy, finance, and programme production, but subject thereto the Council has direct responsibility for certain activities with an annual block grant from the B.B.C., and its own secretariat free from B.B.C. control.

The activities of the Council include: the supervision through committees of programme and pamphlet arrangements and engineering facilities; the control and appointment of the Council staff; the organization of research and of the listening end; including contact with training colleges, associations of teachers and outside bodies.

The Central Council for School Broadcasting consists of 51 members representing the major educational bodies both of administration and teachers, and persons nominated by the B.B.C. The Council exercises its powers of supervision of programme and pamphlet arrangements through a series of programme sub-committees coordinated by the Executive Committee of the Council, which determines the main lines of policy. Each programme sub-committee consists of a member of the Council, one or more specialists, one of H.M. Inspectors, and a number of teachers from schools of different types, who form the majority of each committee. The design of the broadcast courses is, therefore, in the control of persons who are in close touch with the schools.

To complete this picture of the machinery concerned with the B.B.C.'s educational work, reference should be made to adult education upon which, since the earliest days of broadcasting, the Corporation has had an organization comparable with that concerned with school broadcasting. In 1937 the term of office of the Advisory Committee on Broadcast Adult Education expired, and the B.B.C. set up the Central Committee for Group Listening with a measure of independence, financial arrangements, and listening-end responsibilities similar to those granted to the Central Council for School Broadcasting, with the two bodies closely linked together through a common secretariat both in London and in the Regions. Whereas, however, the Central Council for School Broadcasting has power to supervise the Schools programme arrangements, the Central Committee has only power to make recommendations to the B.B.C.'s Talks Advisory Committee on the talks broadcast, with the needs of listening groups as well as of general listeners in mind.

Some Figures

The extent of the service of broadcasts to schools can be seen from a few figures. The programme planned for 1939-40 provided for 26 separate series in the national programme for schools in England and Wales. The programmes occupy broadcasting time for about 10 hours a week for 30 weeks in the year. The programmes are graded from programmes suitable for infants to those designed for pupils in the Sixth Forms of secondary schools.

The extent to which the service is used is seen from the following figures of the number of listening schools in England and Wales at the end of the Summer Term in the years given.

Year	Number of Schools
1935	3,708
1936	5,126
1937	6,890
1938	8,543
1939	9,953

In the last school year, 1938-39, more schools (6,306) listened to the travel talks than to any other series. Other series with particularly large audiences were nature study (5,700 schools), world history (4,647 schools), British history (4,271 schools) and senior geography (4,187 schools).

The extent to which educational bodies, generally, take an interest in school broadcasting is considerable. All the leading educational organizations are represented on the Central Council, and their interest in school broadcasting does not end with appointing representatives. The Board of Education, for instance, has recognized the importance of school broadcasting, and has co-operated in many ways with the work of the Council. The position in the training colleges varies, of course, from one college to another, but most of them are giving their students some instruction in the use of school broadcasts, and many of the colleges have an arrangement by which they receive a lecture demonstration every year or two years from one of the Council's officials. Some training colleges are also encouraging their students to undertake research either singly or collectively on the subject.

The School Programmes

The actual programmes broadcast to schools are the result of the co-operative work of the Council and the B.B.C. The Executive Committee of the Council allots sections of the programme time made available by the B.B.C. to programme sub-committees which deal respectively with English, history, science, geography, modern languages, music, special broadcasts for secondary and for rural schools. The planning of the programme is handled by these sub-committees; its production is the responsibility of the B.B.C. There is constant reference between the two bodies in both activities, and there appears to be no danger of the division of function developing into a tiresome departmentalism.

Each programme sub-committee normally meets three times in the year. At each meeting the committee hears the transmission of one of the broadcasts for which it is responsible, and criticizes it in detail from the point of view of speed, vocabulary, suitability for age range, amount of material, and the like. The committee examines pamphlets published for use by children during the broadcasts, and comments on the comparative popularity of the pictures, their suitability, their size and usefulness, the synopses, and the suggestions for follow-up work. At each meeting the committee receives from the Council's Senior Education Assistant a résumé of the reports from schools and from the Council's Education Officers on the broadcasts since the last meeting. This résumé is discussed in the light of reports from members of the committee, most of whom, being teachers, have been able to see for themselves how the broadcasts were received in the classroom. Finally, the committee plans the series to be broadcast in the forthcoming year, drawing on the experience of its own members and on reports received from various educational sources. The advice of the officials of both the Council and the B.B.C. is also available. The committee can call upon the former for information regarding listening schools and school practice, and upon the latter for advice on the possibility of converting into suitable broadcasting form the proposals they are considering. It is unusual for a committee to plan details for a series itself. More often it recommends that an outside expert or group of experts should be invited to prepare a series on the general lines which it has laid down. The programme sub-committee gives preliminary consideration to the programmes which it is planning as far as sixteen months before the series are due to begin. It makes more definite recommendations nearly a year in advance, and approves the programme in its detailed form six months before the beginning of the autumn term; so that the programme for the whole year may be published in June, thus giving teachers ample time in which to plan their work for the coming year in relation to the broadcasts.

It is the duty of the Schools Department of the B.B.C. to provide programmes on the lines so decided by the various programme sub-committees of the Council. The Department consists of the director of school broadcasts and a staff of programme officials and producers. The programme official has to seek suitable broadcasters, prepare material for illustrated pamphlets for use with the broadcasts, and arrange for the writing of scripts. He has to edit the scripts when they arrive, rehearse the broadcasts, see that transmissions go off smoothly, and afterwards take to heart the reports he receives from his colleagues, from officials of the Central Council, from schools, and from the summing-up at the meeting of the programme sub-committee concerned.

Finding the Speakers

Possible broadcasters are tracked down through the files of the B.B.C. and through the recommendations of the Council's programme sub-committees, of learned societies, Government departments, universities and training colleges, and the innumerable contacts of the B.B.C. and the Council with the educational public and the public at large.

The programme official must train himself to draw people out and to sum them up very rapidly. He must decide quickly whether or not the proposed broadcaster has the knowledge required, the ability to talk to children in a natural straightforward way, a voice and personality which will come sweeping out of the loudspeaker to take up the attention of every child sitting in the classroom, and a mind of such resource and activity that teachers will be stimulated. Clearly a broadcaster must know and like children of the age he is addressing, but there is no necessity for him to be a teacher. School experience is of great help to the broadcaster, but it is often possible for that lack to be made up by skilful guidance on the part of the programme official who has that experience. It is most happily rare for the Council to have to recommend the rejection of a broadcaster because of his inability to adapt himself to the special requirements of school broadcasting.

Reference has been made to the illustrated pamphlets for use by the children during the broadcasts. They are produced by the Schools and Publications Departments of the B.B.C. in collaboration. Extensive research is carried out in museums and libraries in order to find suitable illustrations, and maps and diagrams are specially drawn and bibliographies are carefully worked out in consultation with the Library Association. In addition, there are issued on certain subjects teachers' leaflets which give more up-to-date information and notes on the educational purpose of each broadcast.

Varieties of Presentation

School broadcasts take many different forms—whole broadcasts devoted to a talk by one man form a very small minority. In this respect they are notably different from the group listening talks. There are talks in which recordings and sound effects are introduced to create atmosphere or to illustrate points; interviews and discussions; broadcasts such as history and those for rural schools, in which a number of dramatic interludes are linked by a narrative. There are English broadcasts of all kinds, from dramatic biographies to performances of modern and Shakespearean plays, from book talks to poetry readings. With music, again, there is a considerable variety—community singing, full orchestral concerts, broadcasts on the instruments of the orchestra, and, for the very young children, broadcasts in "Music and Movement." Recently physical training has been added to the subjects dealt with, and in these broadcasts music is combined with the exercises. A further form of broadcast particularly suited to the medium is the feature programme, which aims at bringing actuality into the classroom by, for example, taking a microphone to a factory or dock, and letting the children listen to all that is going on.

The form of a broadcast is carefully chosen to suit the educational purpose and ages of the children, and even when the form has been decided on there are many pitfalls to be avoided if the broadcast is to be a success. The dramatic interludes have possibly a greater appeal than any other broadcasting form, but care has to be taken to make the point of each interlude stand out clearly and, in particular, in history the time sequence of the interludes in one broadcast has to be made unmistakably clear. In a talk, too, the broadcaster has to be on his guard. He will fail to attract the children unless he uses short, simple sentences, short words, and makes his talk vivid by making it personal, and by the use of suitable illustrations.

The Influence of School Broadcasting

It is early yet to give a full account of the influence of school broadcasting in the schools and in education generally. Naturally the influence varies in different schools according to the use made of wireless—and this use varies from an occasional broadcast turned on as an extra, to regular listening as a normal part of the curriculum. The commonest reasons why schools listen may be summarized as follows: Broadcasts

1. Provide help for teachers in subjects on which they are not specialists.
2. Give variety, new voices and new points of view.
3. Give more reality by bringing the outside world into the classroom.
4. Illustrate class lessons by means of dramatic interludes.
5. Provide up-to-date information on a large number of subjects.
6. Set an example by maintaining a high standard of performance.

Not all schools listen for all these reasons, but there must be few who do not listen for one or some of them, though individual schools no doubt have also other reasons of their own. The influence exercised by the broadcasts depends, therefore, on their success in fulfilling these purposes. The Central Council for School Broadcasting invites and receives comments from a large number of teachers on both the broadcasts and on broadcasting in general. These comments are by no means confined to un-critical expressions of appreciation—in fact they constitute a steady stream of constructive criticism whereby the policy and ideas of both Central Council and B.B.C. have been developed and modified. But taken as a whole the outstanding feature has been the encouragement given to school broadcasting by teachers of all kinds all over the country. From this, and from the rapidly increasing number of teachers using the broadcasts, it may be inferred that there is a strong general feeling that this influence is good, and therefore, presumably, that the purposes outlined above are being fulfilled.

The effects of the use of broadcasts would normally be that:

1. The small schools feel themselves less handicapped by lack of specialist teachers.
2. Education in these schools is less associated in the children's minds with the voices of one or two teachers.
3. Remote schools feel themselves less remote and all schools may feel that their work is more closely in touch with the world outside.
4. Text-book facts are clothed with new meaning, impossible with ordinary classroom resources.
5. Teachers are continually provided with new material and the services of a constant "refresher course."
6. In everything which concerns the spoken word or can be judged by the ear, schools are provided with examples by which they may criticise their own performances, with added opportunities for appreciation.

Effects on the Curriculum

Listed above are the chief of the broad educational effects. It is less easy to generalise about the effects on the curriculum. It is the intention that most broadcasts—and this is widely felt to be desirable—should not be considered an end in themselves. When it is said that a broadcast is not a substitute for but a sup-plement to a class lesson, it is implied that its effectiveness de-pends to a great extent on the use which the class teacher makes of it. This implies again that, in drawing up a time-table to in-clude broadcasts, a teacher must allow for considerably more time than the twenty minutes actually spent in listening. Preparation and follow-up are both involved, and the broadcasts stimulate the children to a host of new interests and activities. In fact the place occupied by a broadcast in the time-table may be anything from one school period to several weeks, during which it may form the centre of a "project."

If taken seriously at all, the broadcast must be fitted into the school syllabus. Ways of doing this are the concern of the individual teacher and depend on the purpose for which he wants his class to listen. Though broadcasts are planned in series, they are not planned to form a syllabus, and it is not likely that a broadcast series as a whole will fit a school syllabus exactly, without modification. The alternatives thus open to the teacher are either to modify his syllabus or to choose those broadcasts which fit into it and leave the others. The broadcasts are so planned that either alternative is practicable, for though there are connecting links and continuous ideas running through a whole series, each broadcast is planned as a separate whole which can be perfectly understood without reference to any of the others. The time-table may determine which of these methods is adopted, or some teachers may feel that their independence is threatened if they alter their own syllabus to suit the broadcast series. On the other hand, others may feel that though their own syllabus must be the first consideration, a good series of broadcasts is no more to be ignored than a good new text-book, and may even sometimes form the foundation of their teaching.

Certain difficulties are inevitable. For instance, the broadcasting terms have to be arranged for the greatest convenience of the greatest number. Schools in different parts of the country take their holidays at such diverse times that it is impossible to arrange broadcasting terms which coincide with the terms of all schools. Again, the material broadcast has to be of general rather than particular interest, and may sometimes be found not exactly to fit individual needs.

A Larger Influence

Nevertheless, in spite of these difficulties, the remarkable growth of interest in school broadcasting and of the use made of it is a sign that the difficulties are small in comparison with the advantages. Schools are brought more into touch with each other and with the outside world; teachers are helped to keep themselves up-to-date; parents often listen at home to the broadcasts and discuss them with their children when they come back to tea—a development which has received too little notice, but which must be creating in many homes a new attitude toward education and a closer link between parents and schools; most important of all, perhaps, the wireless has founded a new university to which everyone can belong, and it is not too much to say that for many children wireless provides, after they leave school, the chief opportunity of continuing their education, in however informal and unsystematic a fashion. In this sense the school broadcasts, as is realized more and more by teachers, are a training for life.

Comparison with Other Aids

It is instructive to compare broadcasting with other mechanical aids such as the cinema, epidiascope, episcope, and gramophone. Broadcasting is perhaps the most independent of these in that it gives the most direct contact between the children and another personality. The teacher knows beforehand what is contained in pictures, books, gramophone records, and, possibly, in films; he chooses those which he thinks suitable and he is secure that they cannot let him down. But he has no such security with a broadcaster, who is actually speaking to the class and giving his talk for the first time. And because the children know that when the wireless is turned on there is a real person actually speaking to them, and because this real person may even say things which the teacher does not agree with, some teachers have felt that the wireless is a rival to them in a sense true of no other mechanical aid; in fact, that the wireless is not so purely mechanical.

This difficulty is not so great in practice as in theory. Not only are the broadcasts carefully prepared and all efforts made to give the schools what they want, but the teacher soon comes to rely on his own experience: if a series of broadcasts has given satisfaction in the past, he can be reasonably confident about the future; if it has not, he can always cease to use it. Broadcasters are not irresponsible, and in most cases the teachers have a good general idea of what is likely to be said and of the use which they will be able to make of it. The more direct personal contact of the wireless has therefore come to be recognized as an asset, for which the occasional slight uncertainty is a small price to pay.

The value of films lies in the visual reality which nothing else but actual sight (and sometimes not even that) can give; the value of the gramophone is its utility and the convenience with which records can be stopped at any point and played over again to illustrate or drive home some special point. The distinctive value of wireless is the contact with people actually speaking and events actually happening at the moment they are heard.

The limitations of broadcasting as an aid to the education of children have been indicated in this chapter, and the above comparison with other mechanical aids suggests that the most important from the point of view of the child is that the spoken word over the radio is "gone with the wind." Moreover, any attempt to use school broadcasts gives rise to many difficult problems for the teacher. His job is not made easier by using them. But the fact that so many teachers give serious thought to these problems and in the main overcome them successfully, is the most tangible recognition that one could hope to have of the contribution that broadcasting can make.

Problems of Reception

From the very earliest days of school broadcasting the B.B.C. realized that the success of their efforts to provide programmes acceptable to schools would depend to a very great extent upon a satisfactory standard of reception. There were very many problems to contend with, not the least being the fact that school broadcasting was not then recognized by education authorities to the extent of providing grants for the purchase of sets. Schools had to raise money for themselves in the best way they could. Commercially made receiving apparatus at that time was not only expensive but it was seldom suitable for use in any but a very small room. Further, the requirements of schools are very different from those of the private listener, since school classrooms and halls are much larger than rooms in private houses, and the acoustics are often none too good. It was very clear, therefore, that some means would have to be found for providing sound technical advice, so that schools might be assured of apparatus giving at least reasonably good reception. Therefore, in the autumn of 1926, the B.B.C. set up a small section of engineers to provide a free advisory service for schools and education authorities. Since the formation of the Central Council in 1929 these activities have been directed by the Council; the men providing the service are still looked upon as B.B.C. engineers, and have to act in accordance with both Council and B.B.C. policy.

During the past ten years there have, of course, been very great improvements in the types of apparatus available. In the early days most schools constructed their own battery-operated sets and these were seldom capable of satisfactory reception under the somewhat difficult conditions prevailing in schools. As time went on, more and more sets became available for operation from the electric-light mains, and improvements were made in the standard of quality of both sets and loudspeakers. At the present time, many ordinary commercial sets provide a sufficient volume of sound for school classrooms and halls, together with the high standard of quality which is necessary. At the same time, a steadily increasing number of authorities are taking responsibility for the whole or part of the cost of installation. Apart from electrically operated sets, battery sets have also improved very considerably, and schools not fitted with electric light can obtain quite good results for classroom purposes with modern equipment of this type.

Technical problems at the listening end are the concern of the Reception Sub-Committee of the Central Council, to which is responsible the Approval (Apparatus) Sub-Committee. The latter examines commercially made sets and issues a list of apparatus suitable for schools. To this sub-committee, radio manufacturers submit receiving sets, radio gramophones and loudspeakers which are carefully tested under classroom conditions. The scheme is open to all radio manufacturers, but their apparatus has to comply with certain technical requirements before it can be submitted. The list is completely revised once a year—that is to say no apparatus is carried forward to the new list unless it has again been submitted for approval. In this way the high standard of quality is maintained, and schools are assured that no apparatus is included in the list which does not conform to the present-day standards of the Committee. Many education authorities have realized that the Council and its Committees go to a good deal of trouble and expense every year in examining apparatus and eliminating that which is unsuitable, and they for their part have insisted that their schools should only install approved equipment.

Since the War

In the above description no account has been taken of the changes in school broadcasting caused by the war. The available evidence suggests that, in spite of the dislocation of school life caused by evacuation and de-evacuation, school broadcasting is being used quite as much in war-time as it was before. The main changes caused by the war were the discontinuance of the pamphlets for children with consequent changes in broadcasting methods, the introduction of one Home Service programme which meant that it was impossible to broadcast to schools over such a long period of the day or on certain of the pre-war subjects, and the extension of the policy of issuing teachers' leaflets. The policy of the Council, which has received wide approval among teachers and others, has been to keep as near to normal as possible.

Wales and Scotland

In the interests of clarity, reference has not been made to the differences in school broadcasting in Wales and Scotland. In Wales, a Committee of the Central Council promotes the development of school broadcasting, and supervises the programme and pamphlet arrangements for the four series for Welsh schools, (one in English and three in Welsh) which are broadcast in the Welsh and North Regional programmes. The series broadcast in the national programme are, of course, suitable for and taken by Welsh schools. In Scotland there is a Scottish Council for School Broadcasting which is responsible to the Central Council. Owing to the differences between the educational systems of England and Scotland a separate programme of school broadcasts is radiated from the Scottish Regional, Aberdeen, and Burghead transmitters. Many of the series of broadcasts are jointly planned for England and Scotland, but the remainder are planned to meet the particular needs of Scottish schools.

European Countries

No reference has been made in this chapter to the volume of school broadcasting which, after the B.B.C.'s example has developed in many European countries. To have included even a summary of its use in these countries would have swollen this Report unreasonably. But these generalisations can safely be made.

(a) No European school broadcasting service is yet as comprehensive or as well-organized as the British, although admirable progress has been made in France and the Scandinavian countries.

(b) Britain again has the advantage of most European countries in that all school broadcasts there are under the control of a single authority.

Moreover, since the B.B.C. is the only broadcasting corporation in Britain, the system has escaped the commercialism with which broadcasting elsewhere is often strongly tinged, and from the other great evil—political propaganda—the school broadcasts are equally free.

Apart from the points mentioned above, two divergent tendencies appear in European systems: one, to use the wireless for direct teaching, as in Switzerland—a centralising policy which leaves the class teacher a subordinate role and tends to standardization; the other, to treat the wireless as a useful aid to the teacher while leaving him the first place, and, instead of competing with him on his own ground, to explore new methods whereby the peculiar advantages of wireless may be most fully used. Compromises, of course, are often made between these two extremes, and different countries and systems vary considerably, but Great Britain for a long time now has been wholeheartedly in favour of the second.

VII ˑ CONDITIONS FOR SUCCESSFUL GROUP LISTENING

THE SUCCESS or failure of group listening everywhere depends primarily upon the stability and efficiency of the organizing machinery. Its success in Great Britain is as notably due to that factor as its comparative failure in other European countries. But there are other conditions of success, and these may be reduced to a minimum of four. In considering these four minima we shall examine the extent to which (a) they have been realized in practice; and (b) the extent to which they are likely to defeat the most careful and conscientious planning.

Group Leadership

The most crucial of all these considerations is the quality of the group leader. His function is to carry on from where the broadcaster leaves off, and no more exacting duty of its kind can be imagined. A mere catalogue of the qualities he should possess adds up to a formidable total. He must be agreeable, tactful, authoritative yet not dictatorial—and these traits of character represent the least that is expected of him. He needs to be familiar enough with the topic under discussion not only to discern the form in which it can best be debated, but also in order to answer questions, resolve obscurities, and generally fill up the gaps which the broadcaster is sure to have left in his preliminary statement. In many ways the leader's task is at least as difficult as that of the broadcaster whose duty, after all, is only that of drawing a rough sketch map of the territory to be covered. The group leader is charged with the harassing business of conducting the party along the route so briefly outlined. On some occasions he must take over the group after a talk which has been a failure—and a really good leader has often been known to redeem such a disaster and to stimulate a really successful discussion in spite of it. Even if the talk has not been a failure it may still have been deficient in debating points, and the leader is expected to improvise a basis of discussion which will keep the group occupied and absorbed for another hour or two.

To this knowledge of the subject a group leader should add a considerable skill in the difficult art of conducting a discussion: he must be adept at coaxing dumb mouths to speak without embarrassing them by importunity or impatience; he must know how to cover an awkward silence or break, yet he must resist that nervous tendency to fill in every gap by an attack of garrulousness; he must know how to restrain without offence the irrelevant member of the group; he must know how to switch over that one-track mind which is always present in any group; he must be fertile in rousing fresh attention when the subject has slid into a momentary decline; he must be lucid and concise in summing-up; above all, he must have the capacity to leave his group feeling that they want more next week. Where are such paragons found? Will they work without reward? Can they be trained to the pitch of perfection that this work seems to demand?

One who has had most to do with group leaders makes this pertinent comment:

The part played by the Group leader is all important. He has to lead, yet to submerge his personality so as to stimulate the maximum activity on the part of the group. To those responsible for organizing adult education he presents a new problem. A man may be an excellent public lecturer, possibly a good tutor, and yet fail as a discussion group leader. The B.B.C. has from the beginning given special attention to the training of leaders and to the holding of schools, national and regional, for this purpose. What are the difficulties of group leading? Many otherwise fluent lecturers cannot adapt themselves to the comparatively small talking-space available for them in the pattern of a good wireless group. I have visited groups where after thirty minutes' solid food from the radio a tutor has then begun to give a further half hour on the broadcast. Others are too readily satisfied if they can get every member of the group to say something, however irrelevant. Many otherwise good discussions are rendered ineffective by the leader's omission of a summing-up which crystallizes in the group's mind the specific points illuminated on that occasion. From my observation I would contend that to lead a discussion well, to keep it tidy—relevant, and yet free from any sense of labouring a few points—is one of the most difficult tasks in adult education at the present time.

Who is best equipped to be a discussion group leader? The man with good academic qualifications, the layman with an absorbing interest in some particular subject, the schoolmaster, or the man who has travelled commercially and has personal experience of lands abroad. To illustrate this point more specifically: in a series of broadcasts on the countryside, such as the B.B.C. often provides in the summer, who should be selected as leader—the man with a degree in botany and natural science, or the enthusiast who has spent forty years of his life learning the lore of Nature at first hand? Provided that the would-be leader is reasonably knowledgeable in the subject of the discussion series, the decisive factor in selection must be his capacity for sharing experience with his fellow human beings. However desirous he may be to inform, he must be capable of listening to others, and even if the opinions he hears advanced appear absurd to him he must handle them with consideration. When in discussing a broadcast on "Spain" a woman member commented: "Well, I'm always glad when I hear they're fighting among themselves—it keeps them from bothering about us," the leader was tempted either to pour ridicule or indignation on such insularity. A moment's thought led to a quiet analysis of the remark, with the result that he sent home an enlightened woman instead of a group member nursing a grudge because she had been laughed at.

In the course of this investigation information has been sought on various points from 141 group leaders in Great Britain. Of that total 87 were either teachers or had had some equivalent academic training. The remaining 54 can be classified as keen amateurs, many of whom had some previous experience in the sort of organized discussion which is characteristic of adult education in Great Britain. They were members of such organizations as the Women's Institutes, Clubs, Y.M.C.A., W.E.A.; and, of these 54, none received any kind of remuneration for their work as group leaders. Of the first category—the 87 persons of some academic training—a minority are paid a few shillings a time by certain local education authorities who wish to encourage group listening. Many groups are very ably led—and not invariably those conducted by the academically trained leader. On the other hand, it is true to say that the mortality rate among listening groups in Great Britain depends more upon deficiencies of leadership than upon any other factor. A superlatively good leader can sometimes so broaden the range of a group that it ceases to be primarily a wireless listening group and becomes a variant of the traditional kind of adult education. But a poor leader can neither get the best out of a wireless discussion nor transform it into any useful kind of substitute. Groups are occasionally found which have no regular leader, and hold firmly to the belief that it is a good thing for each member of the group in turn to preside over the discussion. While there may be much to be said for this from the point of view of training in chairmanship, it does not make for group efficiency.

Those responsible for listening groups have completely recognized that the leader is the decisive factor; and in Great Britain very considerable efforts have been made to improve the quality of leadership. One of the methods followed is to organize—often in collaboration with adult education bodies—training courses of varying duration for group leaders. These courses have rung the changes on so many types and have proved so valuable as to be worth a fairly full description here.

The Week-End School

The week-end school has always been one of the best patronized features of adult education in Great Britain, and it has been an easy matter to use them for the special purposes of group leader training. The organizations co-operating with the B.B.C. have invited to these schools 30 or 40 of their members who have a general aptitude for intellectual leadership and a special interest in educational broadcasting. The usual method of such a school is to divide up the students into groups of about a dozen and to distribute them among various rooms, each of which is fitted with a loudspeaker. A talk is then broadcast from a central point in the building by means of apparatus installed for the purpose, members of the group undertaking leadership of the discussion in turn. A tutor in charge makes criticisms and suggestions on the methods tried during the practice. Much can be learnt in this way in a single week-end, during which four or five group practices can be easily fitted in. The advantage of a week-end school is that it can be attended by busy people who would not normally be able to spare a week—the usual length of a summer school. Only people with a good educational background are likely to derive full benefit from such a week-end school. No study of the subject matter of the talks is possible; the main purpose is to give some idea of group leadership from seeing a group at work. Those attending can return home and work out a technique for themselves. It is true that many less gifted people also get a liking for group listening and may turn out to be excellent organizers of new groups. The great increase in the number of groups since the Central Committee for Group Listening was set up in 1937 has been due more to these week-end schools than to any other single factor. The week-end school has the advantage that all who attend are well known to their organizations and are carefully selected, while this method of recruitment makes them weigh the value of listening groups in terms of their own organization. A further advantage is that on these occasions leaders get an opportunity for direct contact and conversation, not only with the national organizers of group listening, but also with the programme-planning officials of the B.B.C.

The Summer School

Another familiar feature of British adult education is the summer school conducted by university extra-mural departments in conjunction with the W.E.A. These schools usually last a month, during which groups of men and women attend for a week or a fortnight to study more intensively some particular subject with which they have been occupied week by week during the normal twenty-four-week session of adult education. In collaboration with the B.B.C. a few special group leader courses have been included in the programme of two or three selected summer schools.

The schools chosen in 1938 were Harlech and Hereford. At the latter, sixteen students were selected by the Summer School

Committee on the recommendation of tutors in the West Midlands. These attended for a week, joined in with the whole school at the general lectures, and for the rest of the time had a series of group practices, two B.B.C. Education Officers being present to take charge of this side of the work. Beyond the fact that the general lectures linked up with one of the subjects for the autumn broadcasts no tuition in the subject matter of the broadcasts was given. The school was useful, nevertheless, in that it showed that a new activity could be introduced without damaging the tone or altering the character of the school so long as the students were selected by or belonging to the organization concerned.

Harlech went a step further. B.B.C. Education Officers again took charge of the actual training in group listening, but the tutors in the school undertook seminar work and general lectures in the subjects of the autumn broadcasts, and gave guidance on reading. Indeed, the training in leadership was an "extra," so that the work of the students ranked for Government grant like that of the rest of the school. Both experiments were highly encouraging, and they showed a promise of setting up a permanent apparatus for both training leaders and building group listening into the fabric of adult education.

During the summer of 1939 four experiments were conducted, carrying the process of co-operation one stage further. As it turned out, no two schools worked exactly in the same way, but all had common features. In all cases every student was selected by the Summer School Committee; each was deemed capable of leading a group after training; the whole responsibility for tuition, both in the subject matter of the course and in the technique of group leadership, was in the hands of the school tutors. Two broadcast series projected for the autumn of this year—"In Work and Out of It" and "The Colonial Problem"—were offered for study. Durham formed a seminar for the former, Bristol and Oxford for the latter; Harlech ran seminars for both. They all had a common object in ensuring that the maximum assistance should be provided for each leader before the broadcasts started. As soon as the B.B.C. had sanctioned the courses in June, outlines were forwarded to the respective secretaries of the summer schools for use by tutors and students. In their turn, the tutors prepared book lists and recommended suitable reading for students prior to coming to the school. As far as possible, too, the students were chosen because of their tutorial class experience in subjects cognate to the broadcasts. After the week's training they were to return home to carry on further reading and study pending the broadcasts in October. At all schools each student put in the same number of hours of study as those taking other subjects, the group practices being in addition to their normal work. At Durham there were nine students, studying under the director of studies, who was especially interested in the course and took charge himself. At the various group practices, several people led in turn and came in for criticism by the tutor. At Bristol, where tuition was given by the director of extra-mural studies, the fourteen students were divided into two groups for practice in leadership, the Regional B.B.C. Education Officer taking one of them. At Harlech, where over thirty students took the courses, there were two tutors, and further assistance had to be found to take charge of the subdivided groups at their practices.

Probably the most intensive work was done at Oxford, where there were two very experienced tutors for seven students. A great deal of individual attention was given to each student. Each small group had several practices in leadership. People from the rest of the school were invited to each practice and in turn each student led the group. Thus by the end of the week everyone had had at least three or four practices and had seen several others try their hand at leadership. This method ensured, in addition to the experience gained by those taking the course, that most of the sixty or more other students at the school gained first-hand knowledge of listening groups.

In his report on the Harlech School the B.B.C. Education Officer for Wales states:—

It was evident during the week that there was a good deal of intercourse of ideas between the students, and all the eighty students present came into close touch with Listening Groups. Many people came to me during the week who were outside groups to say how impressed they were with the possibilities of this new medium. It must not be thought that the special Group Leaders' classes were in any way isolated from the rest of the School. The whole School was run as a unit: all the students lived together and had a common social life—which was probably one of the most important factors contributing to its success. Group Listening was obviously part and parcel of Adult Education generally.

A word on finance. The Central Committee made grants to each of these schools varying from £20 to £25. In every case the Summer School Committees contributed in bursaries and tutors' services amounts at least equal to the grant received. At Oxford, where the expenses per head are among the heaviest in the country, the Summer School Committee made a very liberal contribution; while at Harlech, the South Wales and Monmouthshire Council of Social Service, the various University Joint Committees, Coleg Harlech, and the W.E.T.U.C. together provided nineteen scholarships out of a total of thirty-five.

Great Britain appears to be the only European country which has thus attempted to provide any kind of training for group leaders. Sweden is strongly conscious of the importance of such guidance, but so far has not carried the idea further than an occasional conference of adult education organizations at which the forthcoming talks programme is considered, and ways of treating it are briefly discussed. The neglect of this kind of training may account, more than anything else, for the slow growth and frequent setbacks of group listening in Europe. The cost of organizing such a scheme, as the British figures show, is relatively small when it is done in conjunction with meetings of the adult education bodies; and for those bodies it offers the attraction of giving some of their selected students a training not only in leading groups but to some extent in the general conduct and management of meetings. On every ground the advantages of such courses of training are evident; and until and unless those countries anxious to develop group listening realize the crucial value of trained Leaders they cannot expect the best results from this device.

The Material for Group Listening

Much of the disputation about the kind of "talk" (or other "Spoken Word" material) best-adapted for group listening proceeds from opposing ideas as to the objective of group listening. Some experienced authorities—as Dr. Sven Wilson of Sweden—hold the view that the role of educational broadcasting is to "awaken" rather than to instruct in the usual sense of that word. Among British broadcasting authorities, too, there are several whose doubts about group listening spring partly from the belief that broadcasting is better adapted to stimulating than to cultivating an interest.

Apart from these views, however, it is clear that both the topic and its treatment must depend upon the kind of effect we are hoping to make upon listening groups. If we judge the groups entirely upon their vigour and readiness of debate, we must remember that such a banal topic as "The Channel Tunnel" or some similar pro-and-con theme might evoke the loudest participation of the greatest number. If debate ability is to be the only criterion, then the lowest common measure of the material might be very low indeed. On the other hand, a group might listen to a talk which left them speechless in a double sense: first, on account of its excellence and interest; second, on account of its treatment of matters on which they lacked sufficient knowledge for debating purposes. Cases are not uncommon in which a listening group has tended to become a reception group; and so engrossed have the members been in listening to learn rather than in listening to argue, that they have willingly suffered this fundamental transformation of their original purpose. The dilemma which confronts those who plan the material for group listening is this: a series which may be admirable from the educational point of view may be indifferent as the raw material of discussion; and a series which may stimulate lively debate may be of little intrinsic value.

Between these two extremes of risk, group listening in Europe has, on the whole, taken a wise middle course—holding, if anything, away from the topic which is merely debatable rather than valuable in itself. The general principles governing the choice of material can be summed up in four propositions evolved by the B.B.C. Listener Research Section after a comparative investigation among a panel of group members and a panel of fireside listeners:

1. A talk which is a success with groups will almost certainly be a success with listeners generally.
2. A talk which is a success with listeners generally is likely to be a success with Groups *only if it contains good material for discussion*.
3. A talk which is a failure with listeners generally will almost certainly be a failure with listening groups.
4. A talk which is a failure with groups will not necessarily be a failure with listeners generally. If it fails with groups because it is unsuitable for discussion, it may still be a success with listeners in their own homes.

One characteristic of those countries in which group listening has been attempted is that they appear less nervous about unsuitable broadcasting topics than do the countries where no such attempts have been made. France and Holland are nervous, even in the ordinary talks programmes, of political themes; Italy and Switzerland are shy of religious themes; and so on. Only in Great Britain and in the Scandinavian countries does this apprehension seem to be absent. The watchword of the programme-planners in Great Britain is "Balanced Controversy" and on the whole it is accepted and appreciated by group listeners as well as by the ordinary listener. With this safeguard the British programmes have achieved a very high degree of genuine controversy. What is more notable is that most speakers seem able, while respecting and expressing views they do not themselves share, to give their own opinion the good healthy colour of bias which alone can make a discussion seem really alive.

The Range of Topics

The following list of group listening series which proved popular among British groups gives some idea of the wide range of topics which they have been permitted to debate.

Colonial Problems
In and Out of Work: a Study of Industrial Britain
The Mediterranean
How They Do It Abroad: a Study of the Positive Achievement of Other Countries, Including Germany, Russia, and Italy
Class: an Inquiry into Social Distribution

Children at School: a Critical Examination of the British Educational System
Town and Country: a Study of Contrasting Cultures
The Pacific
The Changing World
Straight and Crooked Thinking
Art and the Public
A Christian Looks at the World: a Statement of the Christian Position

After considering such a list as this it is reasonable to make two positive assertions:

1. In general, the subjects offered to listening groups in Great Britain cover as wide a range as the subjects offered by the adult educational organizations. During the last few years they have rung the changes on politics, economics, international affairs, psychology, philosophy, art, science, and local government.

2. The problem of the programme-planners is less that of selecting the subject than of deciding which of its elements will make a good basis of discussion. Treatment appears to be more vital a concern than topic; and in this respect, no matter how much time and thought a committee may give to preparing the schedule of treatment, the final success or failure is in the hands of the chosen broadcaster.

One of the few dogmatic declarations which may be permitted on this matter is that the choice of the broadcaster is even more decisive than that of the topic assigned to him and his treatment of it. The wisest programme-planning committee considers simultaneously the theme and the man capable of putting it over. His name is far from legion.

The Procedure of Selection

Elaborate machinery has been evolved to select the topics and the treatment which listening groups are most likely to prefer. In Great Britain, for example, the administering body for group listening (The Central Committee for Group Listening)[1] has set up a programme sub-committee, made up partly of its own members and partly of persons who have special experience in plan-

[1] See also Chapter II.

ning talks for adult audiences. To this sub-committee suggestions come from many sources—from individual groups (via the B.B.C.'s Education Officers), from Area Councils for Group Listening, from the B.B.C. Talks Department—and from this volume of opinion and experience proposals are made for definite series to the Central Committee for Group Listening, which considers, modifies, or adopts these proposals. At all the consultations representatives of the B.B.C. Talks Department are present; and after this careful process of sifting, agreed proposals are passed on for the consideration of the B.B.C. Talks Department. The adjudication of proposals is so carefully done that the B.B.C. as a rule is willing to adopt the final list as it stands; and at every step the utmost care is taken to ensure that the final choice represents the consumers' demands and special needs. A similar concern is evident in the no less elaborate planning of group listening talks in Sweden and Norway and Finland.

Types of Presentation

So far in this account of the material broadcast for listening groups the word "talk" has been used. As a matter of fact, group listening has advanced far beyond the limits of the straight talk. In Great Britain it is now rare for one speaker to monopolise the microphone so far as listening groups are concerned. One favourite device is for the main speaker—sometimes called "editor" or "interlocutor"—to bring to the microphone another expert of his own rank and to question him on some particular aspect of the topic. A variant of this method is for the interlocutor to call up two or three representatives of the man in the street who proceed to put to him the problems which harass the ordinary man. These and similar devices all aim to bring a greater sense of reality and immediacy into the discussions. They aim equally at offsetting the inevitable bias of the principal speaker, and in this way they certainly succeed in widening the basis of discussion for the listening group at the other end.

Further than this Great Britain has not hitherto gone. Feature programmes (e.g., of history and social life) are not regarded as material for listening groups, on the assumption that the amount of dramatization in such programmes may blur the plain issues of debate: a supposition which is far from convincing. Nor has Great Britain emulated the Swedish practice of discussing plays— a group listening variant which is very popular in Sweden, where the method is to publish a classical play as a pamphlet, along with a "study-letter" bringing out the special features and problems of the play. The play is then broadcast, and, by that time, listening groups are well equipped to proceed to the detailed discussion of its ideas and its technique. Sweden, again, has not been deterred from using feature programmes as material for group listening. Among the topics so treated have been chronicle plays, depicting, for instance, the story of Swedish emigration to the United States. An expert historian provides the "letter-press," so to speak, in the form of a straight talk; then the Features Department uses its dramatic resources to provide the "illustrations." This imaginative method has been very successful with listening groups, and according to many reports has produced livelier (and no less relevant) discussions than emerge from the plain talk or discussion.

Choice of Speakers

The choice of the broadcaster is in most particulars inseparable from the choice of the topic: and programme-planning committees for the most part realize this relation. Up to a point the specifications for a good broadcaster to groups are the same as those for any other broadcaster of the spoken word—"microphone personality," agreeable voice, willingness to be "produced," and so on. But for broadcasters to groups extra qualities are necessary. For the most part they are the qualities of a good tutor in adult education: the capacity to simplify without distortion; the ability to discern and bring out points for discussion; the undogmatic yet authoritative approach; the power to infect the audience with zeal and curiosity. Now these are qualities of the teacher rather than of the "expert"; and from this distinctiveness of roles proceeds yet another of the dilemmas which have to be resolved by the organizers of group listening. One of the stock arguments in favour of this kind of wireless education is that it can give groups contact with experts who would otherwise be inaccessible to them. The villager in Lincolnshire can imbibe the wisdom of an Oxford professor or a Whitehall notability. But the sad truth is that such eminent authorities are frequently bad broadcasters in general and particularly bad broadcasters for listening group purposes. Bitter experience has taught the programme-planner that, more often than not, celebrity value must be foregone in favour of expositor value; and the consequence has been that broadcasters for group listening have had to be sought in humbler yet more capable ranks. Thus in Great Britain several of the most successful broadcasters to the groups have been men already engaged in adult education, or men with an aptitude for broadcast exposition but no publicity value. From the educational point of view this is no matter, of course, for regret, but it leads to two implications of some importance:

(a) that group listening must soft pedal its claim to bring the super-expert within reach of the ordinary citizen; and

(b) that group listening must look for its leaders—as well as for its audience—within the existing ranks of adult education.

On the whole both these considerations have become accepted in Europe; they are set down here partly to recall the fact that, in the early days, they were not so plainly recognized, and partly as a caution to those countries which may be led to embark on extended group listening through wrong reasons and illusory expectations.

The Importance of Reception

If group listening is to substantiate its claims to be the least formal, most easy-going variety of adult education, it must seek a setting which conforms with that intention. The possibilities of finding the right kind of premises vary a good deal from country to country. In Czecho-Slovakia, for instance, most listening groups (i.e., a quarter of them) used to meet in restaurants, which suggests a happy combination of convivial and intellectual interests. This is one way in which local customs or local institutions may affect the issue. In such countries as Great Britain, where the restaurant is an eating place rather than a social centre, this happy choice is completely unknown.

Opinions differ—as they must—on the question of the ideal meeting place. The private house is claimed by some to be unsurpassable for the purpose—but obviously it all depends on the particular house. At its best, no doubt, a private house could provide that feeling of comfort and sociability which encourages many people to open their minds and their mouths. In Great Britain 25 percent of the listening groups meet in private houses. On the other hand, 36 percent meet on school premises, 25 percent on club premises, 7 percent in public libraries and a tiny percentage in church premises.

These figures really mean very little. If they suggest anything, it is that listening groups are not difficult to please in the matter of accommodation. Very often they are willing to take anything that is available. Many groups meet in private houses, for instance, not from choice but because they cannot secure the loan or the hire of a room elsewhere. Many groups meet in schools because, as one witness put it, "We like to be on neutral ground: if we are guests in the Leader's house, or even in each other's we might feel too polite to argue." Other witnesses said they preferred school buildings because these offered fewer temptations to small talk than a private house and a fire and cups of coffee: a Spartan attitude, perhaps, but one not without its point.

There are "suitable" schools, "suitable" clubs, "suitable" homes, and so on. The deliberate choice of any category has little to commend it, and in practice it has been found that groups have no strenuous or decisive views on the matter. A modicum of comfort and reasonable proximity to their homes appear to be their main considerations—as they are, indeed, in every other form of adult education.

All who have had any part in the administration of group listening agree that indifferent radio sets and imperfect tuning are far too frequent.[2] The broadcasting organizations themselves are, for the most part, powerless to ensure good reception, since the receiving sets are invariably supplied by the bodies responsible for arranging the groups. But the broadcasting organizations all do what they can to insist upon the necessity of good reception. The B.B.C., for instance, circulates a leaflet to groups, in which the do's and dont's are clearly set out; and it arranges

2 See also pp. 236-37.

for its engineers to advise and inspect in any case of peculiar difficulty.

Even apparently minor factors can make all the difference to the enjoyment and understanding of a talk—it is a fact, for example, that group listeners seem to grasp less of what is being said if the radio set is placed behind them, or sideways to them, or at an abnormal height from the ground. Anything which causes strain or irritation seems to be a bigger handicap to group listening than to individual listening, for strain is, in a sense, a contagion.

EPILOGUE

THE FUTURE of discussion groups depends on the future of a belief in the value of discussion itself. British radio may, by virtue of the prestige which it enjoys, do something to encourage a belief in the value of critical listening by continuing to provide the material for such listening; but a public service like radio will find it difficult to maintain that attitude, at any rate in any aggressive form, so long as the country is engaged in a war in which propaganda has got to play a very disagreeable part. The primary responsibility for maintaining a belief in the value of critical listening and discussion groups will fall on those charged with leadership in education and the degree to which it becomes possible for the thought of free men and women to be translated into personal and social action.

In the post-war world will free discussion be a popular pastime, and if so, on what? On art it might be comparatively easy, but what about politics, international affairs, economics, social sciences, philosophy? Will men yearn for action rather than for thought, as the tendency has already been in the latter part of the 'thirties? The demand for discussion has perhaps increased, but the number of those who are tired of discussion has increased at least as fast. If the desire to discuss survives the war, and if at the end of it there is still freedom to do so, will discussion lead anywhere except to "the good life"? Can it have any practical significance for the majority of those taking part? That depends on something much deeper than the power of radio. It depends upon whether there is a sense of common purpose, so that the speculation of little men may seem to bear some relationship to what their responsible leaders are doing. If there is that common purpose—and at present there is little sign of it in the democracies of Western Europe—there is surely a future for radio discussion groups, which should not be regarded merely as a suitable form of educational device for civilization's Second Eleven. It is essential that there should be scholars in the community, but discussion groups should not aim at producing scholars of the second rank or, indeed, of any rank at all. The savour of democratic life does not depend on large numbers of bookish people, but on a vast number of citizens ready to face arguments and facts not necessarily agreeable to them, and in making up their minds not on the basis of wide reading but on material as presented to them. Issues are already complicated enough and are not likely to get any simpler in the near future. The democratic peoples of the world will never understand contemporary problems in any great detail. The mass is too overwhelming. The future of democracy depends on whether its leaders can simplify issues honestly yet clearly enough for the peoples to come to responsible decisions. Radio talks for discussion groups ought to provide endless practice in the honest and objective simplification of great issues. It will be upon the capacity of democratic peoples to respond critically to that sort of presentation of all issues that their future depends—particularly in a continent in which democracy will possibly be enervated rather than exhilarated by the desperate struggle to preserve itself from totalitarian tyranny.

INDEX

INDEX

Bei Fragen zur Produktsicherheit wenden Sie sich bitte an:
If you have any questions regarding product safety,
please contact:

Walter de Gruyter GmbH
Genthiner Straße 13
10785 Berlin
productsafety@degruyterbrill.com